Q 174 .S233 1995

AGE 1568

DATE DUE

Library Store   Peel Off Pressure Sensitive

WHAT EVERY ENGINEER SHOULD KNOW ABOUT CONCURRENT ENGINEERING

WHAT EVERY ENGINEER SHOULD KNOW

A Series

Editor

William H. Middendorf

*Department of Electrical and Computer Engineering
University of Cincinnati
Cincinnati, Ohio*

1. What Every Engineer Should Know About Patents, *William G. Konold, Bruce Tittel, Donald F. Frei, and David S. Stallard*
2. What Every Engineer Should Know About Product Liability, *James F. Thorpe and William H. Middendorf*
3. What Every Engineer Should Know About Microcomputers: Hardware/Software Design, A Step-by-Step Example, *William S. Bennett and Carl F. Evert, Jr.*
4. What Every Engineer Should Know About Economic Decision Analysis, *Dean S. Shupe*
5. What Every Engineer Should Know About Human Resources Management, *Desmond D. Martin and Richard L. Shell*
6. What Every Engineer Should Know About Manufacturing Cost Estimating, *Eric M. Malstrom*
7. What Every Engineer Should Know About Inventing, *William H. Middendorf*
8. What Every Engineer Should Know About Technology Transfer and Innovation, *Louis N. Mogavero and Robert S. Shane*
9. What Every Engineer Should Know About Project Management, *Arnold M. Ruskin and W. Eugene Estes*
10. What Every Engineer Should Know About Computer-Aided Design and Computer-Aided Manufacturing: The CAD/CAM Revolution, *John K. Krouse*
11. What Every Engineer Should Know About Robots, *Maurice I. Zeldman*
12. What Every Engineer Should Know About Microcomputer Systems Design and Debugging, *Bill Wray and Bill Crawford*
13. What Every Engineer Should Know About Engineering Information Resources, *Margaret T. Schenk and James K. Webster*
14. What Every Engineer Should Know About Microcomputer Program Design, *Keith R. Wehmeyer*
15. What Every Engineer Should Know About Computer Modeling and Simulation, *Don M. Ingels*

16. What Every Engineer Should Know About Engineering Workstations, *Justin E. Harlow III*
17. What Every Engineer Should Know About Practical CAD/CAM Applications, *John Stark*
18. What Every Engineer Should Know About Threaded Fasteners: Materials and Design, *Alexander Blake*
19. What Every Engineer Should Know About Data Communications, *Carl Stephen Clifton*
20. What Every Engineer Should Know About Material and Component Failure, Failure Analysis, and Litigation, *Lawrence E. Murr*
21. What Every Engineer Should Know About Corrosion, *Philip Schweitzer*
22. What Every Engineer Should Know About Lasers, *D. C. Winburn*
23. What Every Engineer Should Know About Finite Element Analysis, *edited by John R. Brauer*
24. What Every Engineer Should Know About Patents: Second Edition, *William G. Konold, Bruce Tittel, Donald F. Frei, and David S. Stallard*
25. What Every Engineer Should Know About Electronic Communications Systems, *L. R. McKay*
26. What Every Engineer Should Know About Quality Control, *Thomas Pyzdek*
27. What Every Engineer Should Know About Microcomputers: Hardware/Software Design, A Step-by-Step Example. Second Edition, Revised and Expanded, *William S. Bennett, Carl F. Evert, and Leslie C. Lander*
28. What Every Engineer Should Know About Ceramics, *Solomon Musikant*
29. What Every Engineer Should Know About Developing Plastics Products, *Bruce C. Wendle*
30. What Every Engineer Should Know About Reliability and Risk Analysis, *M. Modarres*
31. What Every Engineer Should Know About Finite Element Analysis: Second Edition, Revised and Expanded, *edited by John R. Brauer*
32. What Every Engineer Should Know About Accounting and Finance, *Jae K. Shim and Norman Henteleff*
33. What Every Engineer Should Know About Project Management: Second Edition, Revised and Expanded, *Arnold M. Ruskin and W. Eugene Estes*
34. What Every Engineer Should Know About Concurrent Engineering, *Thomas A. Salomone*

ADDITIONAL VOLUMES IN PREPARATION

WHAT EVERY ENGINEER SHOULD KNOW ABOUT CONCURRENT ENGINEERING

Thomas A. Salomone
Manager, Design Business Group
Digital Equipment Corporation
Maynard, Massachusetts

Marcel Dekker, Inc.   New York • Basel • Hong Kong

Library of Congress Cataloging-in-Publication Data

Salomone, Thomas A.
  What every engineer should know about concurrent engineering / Thomas A. Salomone.
    p. cm. — (What every engineer should know; v. 34)
  Includes bibliographical references and index.
  ISBN 0-8247-9578-4 (hard cover: alk. paper)
  1. Engineering design. 2. Concurrent engineering. I. Title. II. Series.
  TA174.S238   1995
  658.5'75—dc20   95-13635
                                                                                    CIP

The publisher offers discounts on this book when ordered in bulk quantities. For more information, write to Special Sales/Professional Marketing at the address below.

This book is printed on acid-free paper.

Copyright © 1995 by Marcel Dekker, Inc. All Rights Reserved.

Neither this book nor any part may be reproduced or transmitted in any form or by any means, electronic or mechanical, including photocopying, microfilming, and recording, or by any information storage and retrieval system, without permission in writing from the publisher.

Marcel Dekker, Inc.
270 Madison Avenue, New York, New York 10016

Current printing (last digit):
10 9 8 7 6 5 4 3 2 1

PRINTED IN THE UNITED STATES OF AMERICA

To my wife of 23 years, Kathleen, and our five children, Anita, Eileen, Laura, Alanna and Joseph, who have helped me along the road less traveled.

Preface

The increased globalization of industry is causing an acceleration in the pace of product change. It is realigning the competitive positioning of both major companies and smaller ones. This trend is forcing engineers and engineering managers to respond with products that have increasingly lower costs, better quality and shorter development times.

This trend is felt in every discrete manufacturing company in the world. The effectiveness with which engineers and their managers can respond to these trends will ultimately mean the success or failure of their company, their jobs, and even their country's ability to compete in a global economy.

As a result, engineering managers are examining their basic product development methodologies. Many are turning to a concurrent development approach to product, process and market positioning. Initial results by most who have tried it show some dramatic improvements for specific products. These improvements are often shown as product cost reduction, development time reduction, engineering error avoidance and, occasionally, improved market penetration. Because of these success stories, concurrent engineering is spreading from company to company, industry to industry, and country to country.

The trial and error approach is the most common way companies and development teams are using to convert their development process. Some learning occurs through press articles, trade show presentations, society meetings, and networking through the technical communities. This learning is piecemeal at best. As a result many companies are not achieving the full benefits and competitive advantages of a well-planned, well-understood implementation procedure.

At the core of this problem is the lack of recognition of the fundamental changes needed to achieve sustained success in concurrent engineering. Compounding this problem is the lack of well thought out "how to" books, articles, or information. Most of what exists in the marketplace today is too visionary for engineers and development teams to implement, or too specific, providing only a small piece of the solution.

This book provides a guide for engineers on the subject of concurrent engineering, identifying the fundamentals of parallel development of products, processes, and market positioning. Each team member on a new product team will find specific areas that fit his or her expertise. This book covers teams, tools and techniques from the beginning stages of concept development through production start-up. Engineers and engineering teams will learn how to gauge and ensure effective competitive positioning of their products, how to design products to minimize costs, and how to ensure a high quality product, and how to do all this within a rapid time frame. This book will help engineers learn the latest techniques and tools and their appropriate applications. Engineers can benefit from the experiences of the real life examples presented and apply them to their own implementation projects.

Team structures, roles, and methodology are treated from a practical and somewhat unique standpoint based upon my experience in managing a major new product function with over 68 new product teams.

The overview process of concurrent engineering is a recurring theme used throughout the book, especially in the linking of process steps with the use of developmental tools and techniques. This theme highlights the changes required at the concept phase--one of the least understood stages of design--and identifies techniques to help carry this understanding throughout the remaining stages of development. The overview process presented is one that I have used as an engineer and engineering manager, taught for several years, presented to many distinguished organizations and used extensively as a basis for consulting work with several companies. It has been very helpful for many kinds of companies, from aerospace to automotive to electronics to discrete manufacturing, for both large and small companies. This proven process will help readers better understand their own implementations.

What Every Engineer Should Know About Concurrent Engineering includes flow charts and methodologies that engineers can use to achieve success in their product designs. These charts, pictorials, matrices and flow diagrams help to illustrate the content and simplify the explanations. The book covers methodologies such as Quality Function Deployment, Customer Focused Design, Design for Manufacturing, Cost Driven Design, Design for X, Taguchi's Robust Design, Rapid Prototyping, Simultaneous Development, Market Focusing, and others used in concurrent engineering. Cost Driven Design is a process I have developed and used successfully. It is published here for the first time. Also included is a discussion and illustration of tool structures, such as client/server CAD, project teams, management structures, design phase processes, design reviews, and enabling technologies.

This book is intended for any engineers involved in new product development or who have an interest in this topic, especially those required to use concurrent engineering teams, tools, or techniques. This includes all new product team members as well as their managers and associates within the company. In addition, professors and students who need a practical approach to the latest techniques and methods in use in industry today will find this book very useful, as will developers of software products that support the engineering process.

I have structured this book as a "how to" guide to implement concurrent engineering in industry. Thus, those individuals who are in the midst of developing products and are struggling with the new techniques of concurrent engineering can use it as an analysis tool to determine areas where they are doing well and those that need correction.

I hope you find this book enlightening, educational, and entertaining. If it succeeds in transferring a part of the design process knowledge I've acquired over the years, it has achieved its goal.

Thomas A. Salomone

Contents

Preface		*v*
1.	Concurrent Engineering Background	1
2.	Concurrent Engineering, Fundamentally Changing The Overall Process	32
3.	How To Form A Concurrent Engineering Team	62
4.	Selection Of Key Techniques And Methodologies	100
5.	Selection Of Tools	147
6.	Market Focus Your Design	179

7.	Developing Cost Sensitive Products	191
8.	Quality Focused Design	212
9.	Development Time Management	223
10.	Fitting The Pieces Together	236
	References	*247*
	Index	*251*

Chapter 1
Concurrent Engineering Background

What Is Concurrent Engineering?

The simplest definition of Concurrent Engineering (CE) is the simultaneous development of product and process. The difficulty of performing the task of concurrent development of product and process is not so obvious in this definition. As one might imagine it is hard, if not impossible, to define the process of making a product before a product design has been created. Many important details are not understood and known. In fact, for much of the Industrial Revolution until the 1980's, the most commonly practiced method was to develop the product and process in a predominantly sequential fashion. This allowed the design engineers to develop the product based upon technology and concepts that they believed to be competitive in the marketplace and adequate for repeatable manufacturing.

Why Didn't The Sequential Process Continue To Work?

There are three primary reasons that caused the design process to evolve into a concurrent process.

1) Rapid Pace Of Technology. Technology was advancing at an ever-increasing pace. Some companies were able to take advantage of the new technologies and convert them into products gaining market share. This put tremendous competitive pressure inside the companies that remained behind. Design groups were being forced by these business pressures to develop products in record time, and to provide competitive advantage to regain the lost market position. Once companies fell behind the technology curve it was difficult and sometimes impossible to catch up. Time-to-market, which is a term used to identify the time between the initial idea to the time the first customer product is shipped, became the competitive strategy and rallying point for many companies. The luxury of long product and process development cycles was doomed.

2) Forced Design Cycle Compression. As engineers became pressured to develop products more rapidly, manufacturing input or marketing input became low or non-existent on the priority list. Product development was all engineers had time to do. Engineers became focused on the management dictated requirements and specifications. Product inputs from other functions that might cause a slip in the design schedule were often ignored. Thus, as marketers learned more about the customers' needs and expectations, and manufacturing engineers learned more about the cost to produce the product and manufacturability issues, few of their recommendations could be incorporated into the design under development. Any accepted recommendations were incorporated only if they did not have an impact on the overall product development schedule.

As a result major mistakes were made. Products missed their target in the marketplace. Designs were unproducible or had much higher manufacturing costs than expected. The finished designs were often delivered significantly later than planned or expected. This was due to the multiple design iterations necessary at the end of the development cycle to correct design deficiencies or inadequacies. Thus, companies learned that simply hurrying the design process was not the answer to overall time reduction in the product development.

3) Emerging Information Technology And Methodologies. New design technologies were emerging to help the development process. The emergence of workstations and personal computers allowed computer-aided design (CAD) and computer-aided engineering (CAE) to become significantly more cost effective and more widely used. As a result CAD and CAE capability improved. CAD was not only a way to eliminate the tedious tasks of accurate drawings but it also provided new capability such as multiple views, three dimensional solid object views, electronic assembly and others. Computer-aided engineering (CAE) tools, especially simulation, were being developed to help analyze products in a more technically robust and detailed way.

Electronic communications emerged as a way to speed up the process. By making the communications electronically written, and logged through electronic mail, individuals could be held accountable for the accuracy and dependability of their input. Additionally, the fundamentals of project management could more easily be tracked. Specifically, the time individuals took to respond to requests, and the time taken to respond to action items, once difficult to pin down after the fact, could now be tracked through the logging of messages in electronic mail routines.

Additionally, several new structured methodologies emerged which resulted in better ways to understand and predict the product functionality, cost or market acceptance. As examples, material and assembly costs were minimized using Design For Manufacturing (DFM) techniques. Quality Function Deployment (QFD) emerged as a better way for the marketing, manufacturing and engineering functions to assess and agree on key product features.

Thus, the information technology and the structured methodology foundation required to reshape the development process into a concurrent engineering process was emerging.

Organizational And Behavioral Factors

Besides the three primary factors, there were organizational and behavioral factors that contributed to the problems associated with sequential design.

Organizational Segmentation Forces Multiple Iterations. Many companies organized their design function in a segmented fashion. As an autonomous organization it required minimal involvement from both marketing and manufacturing in the day-to-day activities. This segmentation lead to the development of organizational boundaries affecting the development cycle. The design function, being measured on completed designs, was anxious to turn the designs over to manufacturing. Manufacturing, being measured on cost efficiency, wanted to make sure new designs were producible and did not have unnecessary costly operations. This lead to a phenomenon called "over-the-wall" syndrome. Here each function passes the drawings or designs back and forth until they achieve functional success. This method in itself developed into a formidable barrier to the implementation of concurrent engineering (See Figure 1).

When the sequential design cycle was forced into compression, engineers working in this environment did not have the time to devote to their functional role and the additional responsibility of understanding the market and the newest trends in manufacturing. They substituted opinions from their manufacturing associates and information from trade journals, sales brochures, and catalogs in place of solid manufacturing capability studies or formalized manufacturing start-up efforts. They used relatively few discussions with their current customer base to solidify market needs and expectations. They allowed their own engineering opinion to outweigh the marketer's view.

Figure 1 Organizational Focused Development. *The traditional method of handing off a design has been referred to as the "over-the-wall" syndrome, where the design organization passes the designs to manufacturing thinking they are complete, manufacturing then passes it back to engineering, attempting to improve producibility of the design. This process could go on for months, frustrating both organizations.*
(The artwork is derived from Lotus SmartPics for Windows. ©1991 Lotus Development Corporation. Lotus and SmartPics are registered trademarks of Lotus Development Corporation)

Additionally, when engineers consulted the manufacturing and marketing functions on specific issues, it was seldom with enough time for the marketing or manufacturing functions to develop thorough and well-supported answers. Thus, opinions given were individual experienced based instead of data based such as those provided by modeling techniques or market research. This supported the problem of inadequate inputs, and gave design engineers the feeling that their views were the only ones they could trust. Thus, because of this attitude, oftentimes even excellent inputs from solid marketing and manufacturing data were ignored.

As a result of these behaviors, products were designed a minimum of three times. Once during the concept phase, when marketing plans and inputs were required for management approval. Once during the development phase when engineers did what they believed was needed oftentimes losing substantive key features and marketing needs from the previous step. The third time when the product started in manufacturing, eliminating detail design errors, and improved the products' ability to be produced economically.

Loss Of Trust In The Design Function. Because of the constant changing and evolution of the product, the full expectations of the marketing and manufacturing functions were rarely achieved. Consider the organizational trust issues when product compromises were made by engineering often without effectively looking at the marketing or manufacturing impacts. As previously mentioned, these compromises often resulted in problems when the product was brought to market. Key features were missing, or were non-competitive and manufacturing was unable to produce the product as inexpensively as engineering had predicted. The impact of these problems took its toll over time. As a result, the functions of engineering, marketing and manufacturing, became very distrustful of one another and viewed each other as a necessary evil to getting their jobs done.

Thus, in summary the design cycle compression caused the problems of a segmented organization to translate into severe

Concurrent Engineering Background

business problems, such as long design cycles, cost overruns, missed market opportunities and lack of cross-organizational trust.

What Are The Key Elements Of Concurrent Engineering?

Concurrent engineering is comprised of three fundamental elements. They are collaboration, information technology, and process (See Figure 2).

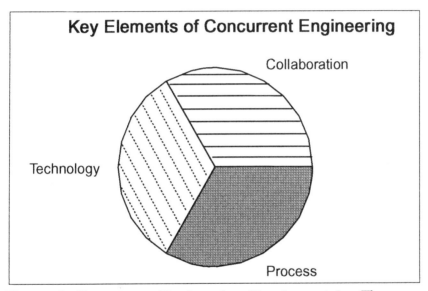

Figure 2 Concurrent Engineering Fundamentals. *There are three significant elements in the implementation of concurrent engineering: collaboration among team members including the virtual team, implementation of information technology as an enabler, and the establishment of formal concurrent processes involving engineering, marketing, and manufacturing.*

Collaboration

Collaboration means to work together or to cooperate with.

In the normal context of concurrent engineering collaboration is usually expressed in terms of teams or teamwork. However, collaboration and teams are important distinctions. Industry teams can operate without deliberate collaboration between the team members. They can simply be a collection of individuals with a common cause. Each person, in order to succeed, follows his own agenda and measurement criteria as established by his individual function. This non-collaborative behavior has lead to many product failures, and is a non-trivial problem inherent to industrial cross-functional teams.

The typical measurements of product success consisting of cost/performance, competitive advantage, delivery on time, are generally not valued to the appropriate level of significance by non-design engineering functions. The result of this is a major conflict for the non-design engineering team member, especially manufacturing and marketing. Since new product introductions often negatively affect traditional business measurements aimed at manufacturing efficiency or controlling marketing expense, the non-design engineering team members can easily cause negative consequences to their own career paths. Any trade-offs committed to by the team member that are good for the product development but negatively affect the organizational metrics have the potential of producing disastrous results for the team member's career. To improve this condition, several industrial studies have pointed to appropriate methods to use to recognize and reward team members and to promote collaborative behaviors. These are discussed in later chapters.

When appropriately used, effective collaboration occurs beyond the team. It includes the full function or organizations within the company, as well as suppliers, customers, consultants, resellers, distributors, and in some cases collaboration with other companies on the development of a single product or technology. Several companies have co-developed very successful products where the strengths of one company are balanced with the strengths of another company. Such is the case when an industrial design firm is used such as Frog Design Inc. Their expertise in understanding appeal, human factors, and functionality in the users environment have lead to many successful and award winning designs, and improved product success in the marketplace.

Thus, in establishing a concurrent engineering environment the infrastructure that drives the behavioral change towards collaborative activities is key. The infrastructure such as reward systems, goal setting, individual recognition, functional measurements, etc. are of major importance in supporting a collaborative environment. Individuals and their willingness to collaborate both within the company and outside the company are a major factor in determining the success of the concurrent engineering process.

Information Technology

Information Technology in the design process is usually discussed as computer-aided design (CAD) and computer-aided engineering (CAE). Additionally and possibly more importantly, information technology has become the foundation enabler that allows concurrent engineering to happen. This is true in several areas of information technology.

1) Simulation technologies for electronic designs, mechanical designs and manufacturing processes are predictive in nature. These predictions allow problems to be anticipated electronically at a stage when corrections are simple and easy to make.

2) It's a communication technology that allows many functions to share information on a regular basis. For example, team members can review data, run their own independent analysis, interpret the results and feed this information back to design. Examples of this include cost models, testing plans and

results, project schedule impacts, manufacturability, customer inputs and market analysis.

3) It provides the communication network that allows the linking of external collaborators. Typically this is done with suppliers who provide feedback on manufacturability and lower cost alternatives that might better suit their processes.

4) Information technology allows the creation of data libraries. These libraries allow the passing of learning and complex accumulations of detailed design data (especially in the area of electronic components) from one product to the next, thus allowing the investment in one product to be leveraged into subsequent products.

5) For some companies their customers are connected electronically. This is a direct method of market analysis and customers can suggest design improvements. This often results in a product competitive advantage. Additionally it allows the development of a better customer relationship through collaborative product development.

6) Information technology provides rapid modeling. Electronic models allow the early review of products, and physical models can be built quickly when the information is transferred electronically to rapid prototyping equipment.

7) Computer networking allows for improved productivity of the entire team. Information technology can be networked in many different ways. At one end of the spectrum there exists mainframe networks, and at the other end of the spectrum there exists standalone workstations or personal computers. The latter is often referred to as an island of automation, with information being passed on diskettes. For the engineering environment a very popular network structure is client/server CAD--clients being the desktop workstation or personal computer, and servers being the database computer "serving" the clients. The advantage of this structure over mainframes and standalone environments is flexibility. It combines the advantages of both structures into a highly effective information network. It allows access to data, by

Concurrent Engineering Background

many individuals contributing to the development of the product as in mainframe structures. However, clients can use their own application programs to work on or analyze the data, or to transfer the data into their own processes. These application programs can be resident on the client or entered via diskette, avoiding the difficulties associated with entering applications at the mainframe level. At the server, revision control can be managed, as well as the systems administration tasks such as information backups and security issues. Client/server groupware tools provide unique applications that allow co-authoring of documents, automated workflow, and the use of multiple applications aimed at engineering productivity improvements.

Thus, information technology provides the enabling foundation for concurrent engineering processes. It provides the automation of design, drafting, and analysis functions. It provides new capabilities that could not exist without it, such as predictive simulations and analysis, and electronic modeling. It provides the communication of information between team members and beyond the team itself. For some companies it's a competitive advantage to be interconnected electronically to their customer. Lastly, through interconnection to production tools and rapid prototyping tools, it provides the means of converting the electronic data files into parts and products.

Process

Process in concurrent engineering involves both the broader level of the full development cycle, as well as the detailed level of specific processes and specific design methodologies. Additionally, it identifies the detailed steps behind the concurrent nature of product and process development. The set of processes to be developed during the concurrent engineering process includes the specific product design process, manufacturing processes, marketing processes, and distribution channel processes.

Creative Development Requires Convergence Of Ideas

The fundamental thought process of concurrent engineering is a process of convergence, where large scale collaborative thinking leads to innovative ideas, market understanding, and product and process knowledge. The result of this convergence of ideas is an engineered product and set of processes for manufacturing and marketing.

The idea of convergence is an important part of concurrent engineering. During the development process, both products and processes are rapidly changing while converging towards a distinct product and a distinct set of processes. During this convergence, learning is being factored into the product and set of processes until a product and set of processes emerge that can produce a highly competitive business result. Convergence is shown in Figure 3.

Convergence helps to keep the level of effort and required learning in synchronization. If convergence happens too quickly in one area the results can be very limiting. For example, early in the development cycle, multiple concepts are needed. However, detailed design done at this stage would be very damaging since it would discourage team creativity and preset many ideas before the proper customer research and process analysis have taken place. Many product teams that have proceeded to develop detail designs too early in the development process have simply ended up throwing away the results when new customer information or new process information was taken into account later in the development cycle. These resets in design are often accompanied with resistance by individuals and their organizations, due to the investment of money and personal creativity. These resets in development can be avoided by recognizing the convergence

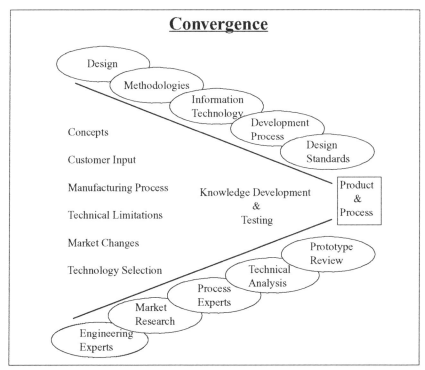

Figure 3 Convergence Of Ideas. *During the development process many inputs are given, many sources of knowledge and information are available, and many tools and methodologies are provided to the engineer. The concurrent engineering process is intended to utilize these multiple sources and inputs in a constructive fashion. In order to do this the engineering team develops and expands its knowledge about the product and the processes, while converging this knowledge into a fixed product and defined processes. During this process the design is constantly changing.*

process of development and the appropriate activities associated with each level of convergence. Engineers should recognize the sequence of acquiring knowledge which requires many inputs, and the constructing of these inputs into product alternatives for the purpose of refining the original inputs or obtaining additional ones. This process occurs and repeats itself until a product emerges.

Marketing Process Development

Marketing process development includes market research, customer testing, feedback to the design team on usability, key feature identification, customer delighters, pricing, product configuration, competitive understanding, competitive positioning, as well as preparation for trade shows, announcements, product demonstrations, and others.

Manufacturing Process Development

Manufacturing process development includes identification of low cost/high quality processes, identification of technology risks, appropriate management of new components and materials, identification of key suppliers, establishment of new component testing and ratings, identification of process qualification methods and identification of designs which can be produced repeatedly with high yields. Establishment of all manufacturing processes also includes fabrication, assembly, testing and quality control, management of all long lead-time parts and tools, development of all prototypes, and engineering support for production operations.

Distribution Channel Development

Distribution channel development includes the negotiations with distributors and product resellers on terms and conditions of the product resales. It also allows the crafting of configurations which will provide business advantages to the product resellers. It identifies the needed product/pricing discount structure so that appropriate product cost and functionality can be addressed by the design team. In some case it includes the identification of collaborative developments and competitive advantages of joint product offerings.

Both marketing process development and manufacturing process development are done largely in parallel with the product development. Additionally, structured methodologies have been developed to provide formal ways of defining inputs from manufacturing and marketing. These methodologies convert these unstructured inputs into forms that can become useful in the product design process.

Collaboration, information technology and concurrent engineering processes are discussed in detail later in this book.

Why Did CE Gain Such Broad Acceptance?

During the late 1980's and early 1990's, companies began to migrate to concurrent engineering as a product development process methodology. Most were driven to the change because of the time-to-market pressures mentioned earlier. Yet many company initiatives haven't had the level of acceptance and success that concurrent engineering has. Why did CE gain such broad acceptance?

The first major reason was the impact on time (See Figure 4). The time required to complete the product development cycle is dramatically reduced using a concurrent engineering approach.

The basic nature of concurrent development means that specific steps can be accomplished in parallel, reducing the total time. Many companies have reported between 30% to 70% reductions in total development time as a result of this conversion. One such company, Chrysler Corporation, learned how effective concurrent engineering could be with the development of Chrysler's Viper. A niche market product, the Viper is a high priced roadster intended to be the successor to the Shelby Cobra.

Chrysler's design team took less than 3 years to take this car from initial concept to volume production, a feat that typically takes 5 plus years to complete.

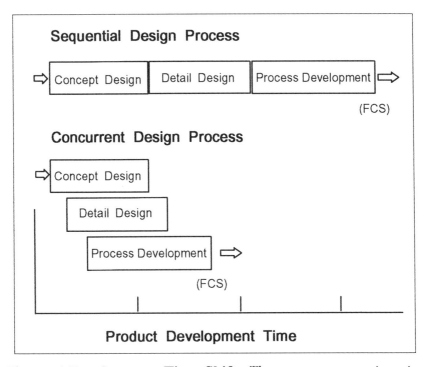

Figure 4 Development Time Shift. *The concurrent engineering process reduces the total time required to take a product from idea to first customer shipment (FCS), using parallel development techniques.*

The second reason is that concurrent development provides a framework for the associated functions of marketing and manufacturing to become involved early in the development process. As a minimum they can begin the development of their supporting processes and in a collaborative environment they can influence the product design for a cost effective/high performance product introduction.

A third reason is concurrent engineering formed a logical platform to justify major expenditures on information technology as important enabling technologies to achieve the benefits associated with concurrent engineering. This helped both the users of the software and hardware as well as the developers of the software and hardware. Engineers were able to get the hardware and software they needed to perform complex design tasks. They become more predictive of the performance and manufacturability of their design. This focus helped the suppliers of this equipment and software to direct their development efforts towards providing concurrent engineering advantages. The result was a series of groupware tools, database architecture, CAD/CAE applications, frameworks, and environments that facilitate the concurrent engineering process.

Thus, for multiple reasons concurrent engineering began to emerge. For individuals and their success within the organization, concurrent engineering was an important advance. CEO's wanted it as a competitive strategy, i.e., to develop many products rapidly to stay ahead of their competition. Marketing wanted it to get products that provided a competitive advantage in order to make selling easier. Manufacturing wanted it as a way to get the products they could cost effectively produce. The product design engineers wanted it as a way to justify new CAD/CAE packages allowing them to be more predictive about their designs in performance and manufacturability. Managers in all functions wanted it as a way to bring teamwork and collaboration into the design process. Thus, for many diverse reasons it rapidly gained broad acceptance as a development strategy.

Should Companies Adopt CE As The Umbrella Strategy?

For manufacturing companies who wish to convert to a formal concurrent engineering environment, should they adopt CE as the main company initiative, i.e., as the umbrella strategy under which other initiatives fall or should this initiative reside elsewhere, for example, within the engineering department?

One drawback of treating CE as an initiative sponsored by an engineering function is that many believe that CE applies in general to relatively few individuals. Those individuals are seen to be within the design and new product organizations of the company. The truth is that full collaborative development occurs only when *all* functions are involved at almost every level. Additionally, each function must have a prolonged commitment to support a concurrent engineering environment.

Effectively implemented, CE will provide the product and process basis of competitiveness for companies, and provide it in a way that can be expanded to include the full product line. One company recently estimated that 30% of its product line will be new products in its next fiscal year. Imagine the effect if most of those products were gaining market share as a result of using the proper CE techniques and methods. Then imagine if this was repeated in each of the following two years.

Thus, treating CE as a major initiative or philosophy of a company is *absolutely critical* to the company's future competitiveness. History has shown that converting to concurrent engineering will be most successful if it is driven at the CEO level. For engineers, engineering managers and the V.P. of engineering, convincing the CEO of their involvement is not that easy. A persuasive case must be made. Thus, consider the following five facts and their potential impact to a company's success. These facts apply to small companies (10 - 15 employees) as well as larger ones (50,000+ engineers).

1) Short Development Cycle Is A Key Success Factor. Getting the product to market to coincide with the anticipated change in customer requirements offers competitive advantages. The earlier to market with unique critical features, the higher the price that can be charged, and the longer the product will remain in the marketplace. Early product availability and aggressive pricing allows market share growth. Additionally, the possibility of establishing standards in the marketplace, or winning a patent, or identifying your company as the innovator in the field, all have significant advantages in gaining customer confidence and market share. In other words, *Shorter development cycles win!*

2) Anticipating The Changing Market. Customer satisfaction is highly dependent on receiving competitive products adequately priced. In order to continually do this, new products must emerge on a frequent basis. The customer's changing expectations must be anticipated early in the design cycle and factored into the product. A competing product which has better appeal to customers' critical needs can dislodge market share, and substantially affect a firm's viability. Concurrent engineering allows this appeal to be developed into the product.

3) Customer Satisfaction. Customers are satisfied when they receive a quality product at competitive prices. Unique features allow additional pricing only when the value is perceived by the customer. Critical unique features to the customer are worth paying more for than features of convenience. The design cycle is the place to resolve these differences, specifically, understanding which features are critical, and spending serious design time addressing these critical features. Additionally, understanding the dollar value of these critical features in the marketplace and pricing the product according to the customer established value.

4) Product Quality As A Design Function. A high percentage (some report as high as 80%) of the product quality is locked in at the time of design. Product quality and reliability are locked in when specific components are chosen, and the design is completed. Only design change can improve quality at the product definition level. Manufacturing, the historical main focus for quality programs, can only insure product quality is not degraded through process variations. Quality in design represents the long term success of any product. Customers, managers, and colleagues rarely forget when something is done wrong or inadequately. Poor design quality in a product has killed many potentially excellent products and/or damaged the reputation of many companies.

5) Profit As A Function Of Design. Product costs are locked in at the time of design. This is especially true for products that are material intensive (See Figure 5). Some estimate this to be as high as 70% of the complete product cost. The cost of materials selected is locked at the time the selections are made, usually very early in the development cycle. The number of components required is just one cost factor dependent on design. Some simple examples of reducing costs early in the design process are: the shrinkage of components through circuit integration, reducing the number of components; or good signal quality in electronic designs reducing the number of filters needed.

In addition, labor content is highly dependent on the processes identified by the design, such as the amount of welding required in sheet metal, or the types of fastening used in assembly. For process industries, the factoring of process capabilities into the design process allows yields to be higher and process equipment to be used at its maximum capacity. For example, designing printed wiring boards to fit cost effectively within the standard sizes of the suppliers raw material sheets, i.e., maximizing the number of boards per sheet of material to minimize wasted material.

Thus, CE as a corporate strategy is one of the few ways in which customer satisfaction, market share penetration, profitability, quality, and long term survival of a company can be achieved. Its techniques and methodologies can focus at the very fabric of product competition and allow the members of the enterprise to understand their contributions to the success of the company.

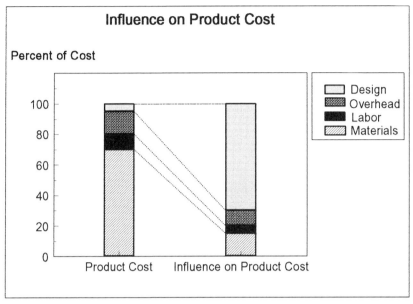

Figure 5 Design's True Impact. *For many companies, 70% of the product costs are controlled by design. Manufacturing has little freedom to improve costs once the design is set.*

Senior Management Must Actively Support The Concurrent Engineering Process

For companies involved in the CE process, the senior management team, including the CEO, must actively support the process. It requires significant change for many companies. New concepts of organization, integration and collaboration are needed. The technology infrastructure must be established and effectively used. Modernization of the supplier and production infrastructure is required. The human component of change is significant, often referred to as corporate culture shock. A substantial amount of training and change in the corporate political infrastructure are usually fundamental requirements.

How Does Senior Management Participate In The CE Process?

Concurrent engineering at its core is a collaborative development process that involves the senior management, a structured development process, a core project team and a specific phased review process. The senior management team commissions projects and project budgets, they establish the overall structured development process, and assign core team members. The team is accountable for the product development. At structured steps within the process the core team reports back to senior management for status updates, advice and help. At each of these steps, funding approval for the next step is partially withheld until the review is successfully passed. The senior management plays a very active role in key product developments. Some examples are; the team may be taking too large a risk on a new technology; the success of major projects may be dependent on key relationships with external suppliers or customers; the positioning of products might affect significant plans for marketing campaigns; the marketplace may be changing more rapidly than originally anticipated, requiring either the expansion of or, as importantly, the cancellation of the development effort. These decisions can only be adequately addressed by the senior management staff.

Expanding The Definition Of Concurrent Engineering

Many individuals and companies use a larger definition of concurrent engineering than is shown in the beginning of the chapter. An important and excellent one was defined by the Institute for Defense Analysis as:

> "Concurrent engineering is a systematic approach to the integrated, concurrent design of products and their related processes, including manufacturing and support. This

Concurrent Engineering Background 23

approach is intended to cause developers, from the outset, to consider all elements of the product life cycle from conception through disposal, including quality, cost, schedule and user requirements."

IDA Report R-338

The Department of Defense and the aerospace industry have been strong proponents of CE. They viewed it as a way to reduce costs, improve quality and shorten development time. The Department of Defense and its major contractors and subcontractors have implemented concurrent engineering in some form with excellent results. The aerospace industry has used concurrent engineering to stimulate and change many of its basic practices of development. For example, the Northrop Corporation successfully implemented team design in the development and implementation of the leadership technologies in the B-2 Stealth Bomber. They viewed concurrent engineering as an element of their total quality management program with roles, responsibilities and reward systems being overhauled in its implementation, and even relocated individuals to better include them in the design team activities.

Another example of concurrent engineering exists within McDonnell Douglas Aerospace. Their effort on the U.S. Space Station, Freedom, has been widely recognized for its success throughout the industry. Their Concurrent Engineering/Integrated Product Definition Team (CE/IPD) comprised of all appropriate disciplines, customers, and suppliers also included an experienced CE mentor who provided team training, guidance, as well as lessons learned from previous activities. The team consistently used CE methodologies. One of these, Design for Experiments allowed system analysts to reduce the number of non-linear rigid body dynamic computations by 65% saving more than 700 hours per analysis model. They also leveraged an impressive implementation of a nationwide network of 350+ UNIX Workstations and over 2,200 PC's. Their software included among others 3D solids modeling, analysis tools, visualization and animation of parts to demonstrate the operation of the Space Station. Their networks allowed simultaneous access by team members in distant locations. Thus, the same files could be viewed at the same time and the details discussed. The results were impressive with a 30% reduction in cycle time, and a 84% reduction in engineering change orders when compared to previous projects without CE. They also achieved 99.4% first time quality. Their design technology implementation won the "Excellence in Computer-Aided Engineering" award in October 1993, awarded by *Computer-Aided Engineering* Magazine. Additionally they were declared the "Best Practice" within the industry by the government-sponsored Best Manufacturing Practices Survey Team.

On the commercial side of the aerospace industry, the Boeing 777 used design-build teams as a major method to include manufacturing and suppliers in the early decisions so that aggressive weight goals could be achieved and the development time tables could be met. Boeing implemented a major new 3D solids computer-aided design (CAD) technology and used electronic assembly as a way of assessing assembly and maintainability on the CAD screen before parts were manufactured. They even developed human figures in CAD to be able to understand realistic sizes and range of motion.

The Department of Defense (DOD) and aerospace industry had a corresponding activity in the commercial high volume manufacturing world. Many of these companies were implementing similar but less structured activities often referred to as Simultaneous Development. These companies had a unique condition when contrasted with the DOD and Aerospace industry. That is, their customer base was exceedingly broader. They did not win project awards, but rather developed products in hopes of acceptance in the marketplace. Therefore, it was difficult to adopt the DOD definition as it stood.

Concurrent Engineering Background

In the commercial volume manufacturing environment, the practice of engineering includes the consideration of the marketplace from the outset. Price and market acceptability are key items of major importance. Additionally, timing is of significant importance in marketplace product positioning. Thus, borrowing from the DOD definition and expanding upon it, a broader definition for the commercial volume manufacturing environment can be defined as:

> "Concurrent engineering is a structured product development process which allows the simultaneous development of market, product, and processes and focuses on market penetration. It causes developers to consider customer requirements, marketplace research, competitive demands, pricing, product trade-offs, the ease of manufacturing processes, product costs, product quality, product environmental issues, regulatory issues, sound engineering practices, and user requirements in the development of products. It provides for enabling information technology. It expands the collaborative nature of development to include marketing, manufacturing, suppliers, selected customers, resellers, consultants, and others as appropriate for both competitive positioning and the effective implementing of the technologies selected."

This is expecting a great deal from the companies (large or small) that implement concurrent engineering, and it is important to note that effective organizations do not leave the above items solely to the teams that are developing the products. Rather, they look at these items over the long term and provide the product team guidance on the issues requiring significant efforts, research, and the development of collaborative relationships. Thus, concurrent engineering isn't done by the "team" alone it is done by the entire organization and the team uses the information, technologies, and research available at the time of development.

There are several organizational competencies that foster and promote concurrent engineering and are addressed in subsequent chapters. For small companies without strong organizational support many of these competencies and information are found through collaborative efforts.

Can Efficiency In Design Be Maintained In The CE Process?

Efficiency in design can be looked at as two separate and distinct items. First, efficiency in design can be looked at as the economic usage of materials selected, and resources consumed to manufacture a product. Second, efficiency in design, can imply the productivity of the engineering team developing the product. Both are important in any design process, especially a concurrent engineering process. Concurrent engineering due to its collaborative nature can degrade both efficiencies if the process is not carefully managed. First, many of the new and contributing team members do not have the technical understanding and the rigorous training that engineers have acquired. They are not fundamentally aware of some of the constraints imposed by materials, geometry, laws of physics, etc. Additionally, there is a good probability that they are excellent at speaking and convincing team members and managers of the merit of their ideas--a potentially dangerous combination of skills for a design team member. Lastly, it is easy to embarrass them due to their unsound technical input, thus causing them not to contribute in a productive manner in subsequent discussions. These are just a few of the conditions that exist and can cause teams to be inefficient, unmotivated, and to potentially end up with products that are not successful.

The method of managing this inefficiency is to put a framework around the team process. This framework provides the overall structure to the development process. This framework includes roles, responsibilities, phases of the development cycle, and includes formal methodologies to insure that proper ideas and inputs are considered. The framework should measure the team in total and individuals for their defined contribution. It should also segment the development process so it can be managed to appropriate schedules and development cost targets. A structured development process provides the link to keeping the efficiencies of the process and product in line with sound business expectations.

Applying Structured Design Processes To Small Companies

A structured design process may seem like a large company approach to development, but it is also the proper approach for small and medium-sized companies. In these companies the managers do a significant amount of the actual design work and product development direction. However, without an agreed too development process much of the work can be wasted, or redone with late inputs, new decisions, etc. A structured design process is a method of avoiding the late design surprises, for example, the technical and supply difficulties brought on late in the development cycle by suppliers; or late changes due to changing market interest in the companies product. In addition, without a structured design process, solid input from the other functions may never be achieved because of time constraints. (This is the too busy to participate syndrome.) The concurrent engineering structure forces a discipline that helps even the smallest companies succeed with their product development.

Many small companies bid for work based upon modifications to existing designs. The project type business also is helped by the concurrent engineering approach of teams, early involvement, communication tools, CAD, customer focused design techniques, and in simplifying the development steps.

Structured development implies a step-by-step sequence of events. For concurrent engineering much of this process happens simultaneously by multiple individuals. To control this process a planned development process is needed. For small companies this book (especially chapter 10) may provide sufficient structure. Larger companies will need their own more detailed structure with organizational roles and ownership, and can be based on the CE process defined in the flow chart in chapter 10.

Small companies have some advantages in implementing CE. By their very nature they are more flexible, allowing them to invest less in physical and organizational structure such as dedicated team members or computer equipment. For example, one small company had their design engineer support the technicians testing the product, develop test procedures, and work on new products. Another company leased extra equipment for the period of development and returned it at the end of the development period.

A Summary Of The Basic Concepts Of CE

Many of the pieces for effective CE implementation have been discussed, others have not and will be discussed in later chapters. As a framework for understanding, it is important to list the elements that fall within a CE development process.

Collaborative Elements

Organizational Structures
    Senior management support structure
    Cross functional teams

Organizational Support
    Marketing trends
    Competitive analysis
    Product benchmarking
    Market research processes

> Collaborative relationships with the "virtual enterprise" of suppliers, resellers, key customers, technology developers, etc.
> Supplier research
> Process technology development
> Design enabling technology development

> Skills
>> Talented and competent engineers and engineering managers, and team members
>> Collaborative willingness and propensity
>> Ability to recognize one's own strengths and weaknesses

Processes

> Overall Processes
>> Structured development process/concurrent development
>> Structured senior management review process
>> Concurrent development of concept, product and process
>> Rapid product development techniques

> Methodologies
>> Concept design methodologies
>> Market target methodologies
>> Product functional development methodologies
>> Quality function deployment (QFD)
>> Cost driven design (CDD)
>> Customer focused design methodologies
>> Structure design analysis
>> Design for manufacturing (DFM)
>> Engineering analysis and simulation

Design of experiments
Robust design
Formal manufacturing principles
Others

Information Technologies

CAD/CAE tools
Client/Server hardware and networking software
Database software
Analysis and simulation applications
Communication tools
Environment/desktop software

Chapter Summary

This chapter provides substantial background information on Concurrent Engineering (CE), and sets the foundation for the subsequent chapters in this book. This chapter identifies the following:

1) The driving forces that caused the emergence of the concurrent engineering process.
2) Three key elements of concurrent engineering were defined and explained: collaboration, information technology, and structured processes.
3) The reasons companies should adopt concurrent engineering as their key competitive strategy were highlighted.
4) Examples of CE implementation were given.
5) A commercial definition of concurrent engineering was

developed. This is an operating definition showing key areas and principles to be followed.

6) Questions were answered concerning the impact on the efficiency of the development process using CE techniques.

7) A listing of key items that comprise concurrent engineering was provided.

Chapter 2

Concurrent Engineering, Fundamentally Changing The Overall Process

What Is Fundamental Change ?

Companies that convert to an effective concurrent engineering environment must undergo change at a very basic and fundamental level. Fundamental is defined as the essential level or foundation. It can be considered the basic principles upon which all other parts are formed. This level of change in an organization takes years before it is fully realized, and it can be derailed at almost any stage of the process. Therefore, a long term plan and a long term commitment by management is needed. This commitment requires investment, organizational change, new skills and enabling technologies. This chapter identifies some of the fundamental changes required for effective conversion. However, without the commitment to fundamental change, companies will not achieve the level of concurrent development

Fundamentally Changing The Process 33

that will be a requirement to stay competitive in the decades ahead.

What Changes Need To Occur ?

Concurrent engineering is the development process which provides a method to minimize life cycle costs (including development costs, product costs, start-up costs, process costs, etc.), maximize quality (including customer input, reliability, output quality), provide competitive advantage, (through competitive features, benchmarks, market research, etc.) and does this in a very short design cycle (short enough to consistently avoid market share loss). These results occur as a direct effect of applying the concurrent engineering process as defined in this book.

Where To Begin?

There are several possibilities for identifying places to start the implementation of concurrent engineering. Some of them are:

1. Putting in place the enabling technologies and processes
2. Structuring teams and letting them define the needs
3. Structuring the financials and allowing that to drive the results.

All of these views have pluses and minuses as a place to start. If one focuses on the logical approach of structuring the enabling process, and then developing products, the risk is that an adequate understanding of concurrent engineering doesn't exist. Therefore, the proper underlining processes cannot be adequately established. If teams become the focus of the change, tremendous cost and expensive risks are undertaken. Since there does not exist a set of overriding management principles, the team is left to determine good and effective spending guidelines. In the view of the author, the right place to start is by structuring the financials over a time horizon, more specifically in the normal budgeting process of an organization. This forces individual functions to begin to understand the change as an organizational issue, a budgeting issue and a role issue. It allows management to define the process financially first, and then hold the team and the functions accountable to achieve the financial metrics, and more importantly the business results.

What Are Some Of The Budgeting Guidelines?

When deciding on a budget, there are some general guidelines that are helpful. First, all functions have a role, although the predominant roles are marketing, engineering and manufacturing. Second, all the work is not done by the concurrent engineering team. The team should be focused on the product development effort. The organization should have separate functions that develop the concurrent engineering environment which is then used by the teams. As companies start into this process, many of the underlining technologies, principles, methodologies, and base understandings will be very inadequate to do an effective job. Therefore, it is important to understand that the first team effort will mostly help the functions understand the key areas in which the process is lacking, and the real benefits of concurrent engineering may not be seen. The second iteration should do much better, and the third iteration will start to fit the mold of an effective CE practice. Subsequent product development efforts will continue to improve the process until the entire organization is structured to support it. Thus, a long term strategy over a series of product developments should be planned and structured at the outset, so realistic expectations can be set for the initial start-up.

Fundamentally Changing The Process 35

Investments For Reduced Development Time And Costs.

One of the key benefits of the CE process is its immediate impact on the product development time and product development costs. This is shown in Figure 6.

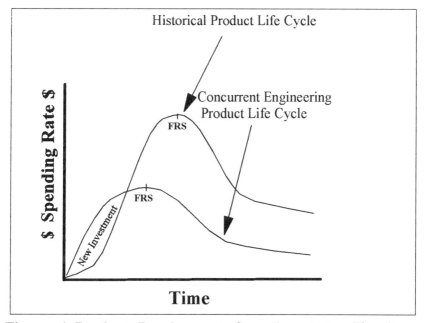

Figure 6 Product Development Spending Rate. *The figure above shows the comparison of a historical development process and a concurrent engineering development process as they relate to the development spending rate. Not only is the time between start-up and the first revenue shipment (FRS) shorter, but the overall development spending over the life cycle of the product is reduced. Significant product changes are avoided. In many cases entire iterations of the development cycle are eliminated. Additionally, substantial cost reduction ideas are included early on.*

Even ineffective approaches to CE have shown substantial improvement in development time and spending. The effect of the additional investments required by the CE process on the overall development process can be seen in the chart in Figure 6. This chart shows that there is a requirement for more investment in the product team early on in the process cycle. This investment funds the participation of the other functions and sometimes suppliers, customers and resellers. It also funds the new activities associated with concurrent engineering. This investment allows these other functions to contribute to the development of the concept.

The purpose of the investment is to maximize the return. Early in the development cycle the design can easily be changed to take into account the inputs of other functions. This improves the overall business success of the product. These changes are very inexpensive to make at this level in the design process. In previous design processes, either these inputs where never incorporated into the design, or they were made at a costly rate later in the development process.

Engineers have for many years been taught the rule of tens. It is ten times more expensive to make design changes at each significant level in the design process. Thus, this rule of thumb suggests that market requirement changes or manufacturing producibility changes found late in the design cycle are very costly oftentimes 100 to 1000 times more expensive to implement than if they were made at the concept level.

In summary, early investment in the development activities really is a pay it now, or pay it at a much more expensive rate later proposition.

Investment For Reduced Product Life Cycle Costs

An important concept is the understanding of the influence design has on the life cycle cost of the product. A simple way to understand this is to examine the relationship between the design process and product cost. The process of design at its essence is to develop knowledge about the design, continually defining and refining this knowledge until a product and the processes needed to market, sell and manufacture are developed. During this process, the cost of the product becomes established. As can be seen in Figure 7, the majority of this cost is determined early in the development process. This is a time when little is really known of the product, the processes, the customer and the market. It is also a time when very little is being spent. This highlights the large leverage design has in setting the business profitability. Investments made to improve this knowledge are the key to success. These investments should be made both with the development team and in the supporting functions, since these functions provide essential information to make the design team successful.

Concurrent Engineering Improves Concept Development

In the concurrent engineering process, concept development is a prime area of change. During the concept phase, customer inputs, market inputs, competitive information, manufacturing process information, business partner information, as well as other information are brought together by the different functions that comprise the development team. New methodologies and processes are available to help the design team do a significantly better job of impacting the business results during this phase of development. As a result of this added effort, the concept phase is longer, and the subsequent phases are substantially shorter.

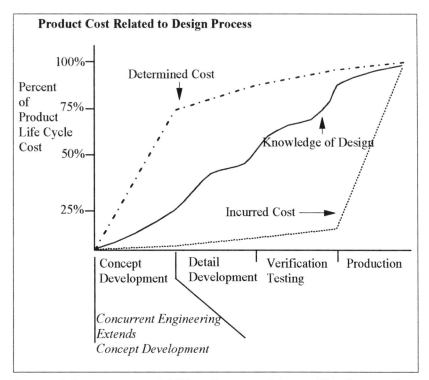

Figure 7 Importance Of The Concept Phase. *The figure above shows the relationship of the development costs (incurred costs), to product costs (determined costs), and relates the level of knowledge about the design to these costs. As can be seen, much of the product cost is determined by the early decisions the design team makes. Historically, these decisions were made by a few individuals before substantial knowledge was developed about the product, the customer and the processes. In addition, these decisions were made before much of the costs were incurred. In the concurrent engineering process, the concept phase is longer and more involved in order to take advantage of the multi-functional inputs from the team and their functions. This allows key product cost decisions to be made with an improved knowledge base about the design and maximizes the design team's leverage on product profitability.* (Graph Courtesy Prof. M. Henderson Ph.D., Arizona State University)

Fundamentally Changing The Process

What's The Detail Behind The Parallel Development Process ?

The development process without a concurrent engineering mindset can still have market assessments, manufacturing inputs, or changes made to the product by the engineering community prior to shipment. These ideas constitute the nature of development. The main difference is in the way inputs are handled, when they are handled, and how they are handled. In the next sections of this chapter the historical development process is contrasted with the concurrent engineering development process in order to highlight the key differences. First, a process overview of product development without the concurrent engineering mindset is given. Secondly, a process overview with a concurrent engineering mindset is shown. This is then followed by a detailed flow for a concurrent engineering process so that one can get a perspective of the overall process and understand the new steps, tools, and methodologies that are involved.

Commercial Product Development Process Without CE

The development process without CE can be seen in the chart in Figure 8. It starts out with market needs being identified. This can happen in several ways: customers start buying competitors products, managers anticipate market change, etc. Once the market needs are understood the product specification is developed and engineering gets started in the design cycle. Next the product is defined, prototypes are made and the product is tested and modified depending on the test results. Once engineering is confident that it will pass testing without substantial change then manufacturing is brought into the picture. Manufacturing identifies the manufacturing process to be used, and the way in which the product will be tested. This normally results in some respecification and redesign on the part of engineering.

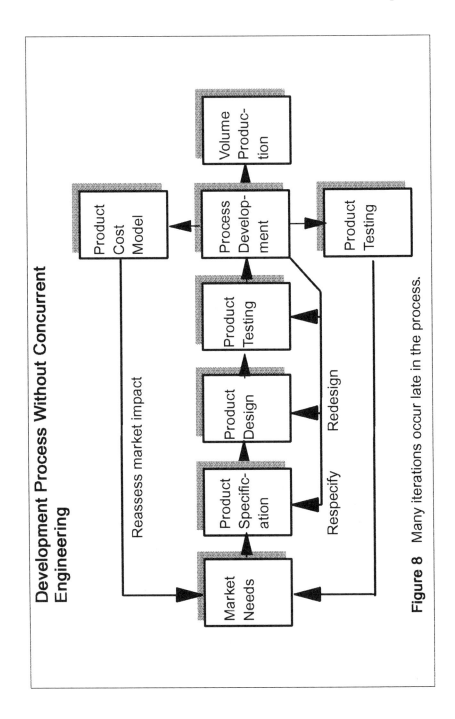

Figure 8 Many iterations occur late in the process.

Fundamentally Changing The Process

At this point, additional prototypes are built and the complete cost estimate for the product is developed. The testing information and the product cost information are now reviewed to see if the cost objectives and functionality objectives are met. In many cases they are not and the cycle is partially repeated. Oftentimes a decision must be made to accept something that is less than what was expected. These missed expectations often have a major effect on marketability and market penetration of the product. Additionally, over the period of the development cycle the market needs have changed, competitive products have been introduced, etc. These changes tend to be factored in at this point in the process oftentimes causing major redesigns in the product. Thus, the historical process has many iterations of the design. As a result, the initial business objectives of the product are often not achieved.

The Development Process With Concurrent Engineering

The development process using concurrent engineering is substantially different. See Figure 9. First, there is ongoing work that feeds the product development process. This work either never got done in the historical process or was developed during the product development process at a great impact to the development schedule. The areas that are developed ahead of the development cycle are:

Process Technology
Design Enabling Technologies
Market Competition Assessments

Process Technology. In the process technology area, research and development is undertaken to define process technologies which will help reduce the cost of producing a product, provide additional functional capability that did not exist

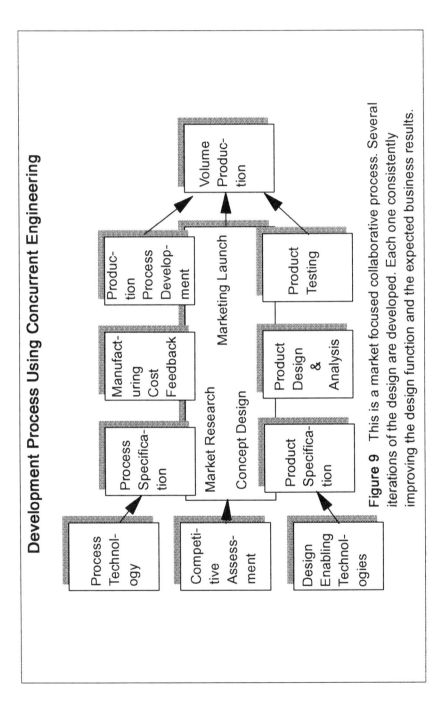

Figure 9 This is a market focused collaborative process. Several iterations of the design are developed. Each one consistently improving the design function and the expected business results.

prior, or launch a new process entirely. Process technology can be factored into the development of the product when the new process is capable of producing products of consistent quality, (even though yields may be very low). Breakthrough technologies and their competitive advantage are achieved by utilizing a process which your competition has not developed. It is a fundamental key to success of many of the major companies.

Design Enabling Technologies. In the area of design enabling technologies there are a two approaches. First, there are unique or patentable ideas which can be converted into competitive products. These ideas allow the development of single or multiple products. Second, there are engineering process technologies such as information technologies, testing methods, and laboratory technologies. These provide competitive advantages when used effectively in the development of a product. By using these technologies an engineer can consistently provide improvements in product functionality, product costs, and product quality, and help provide competitive advantage in the marketplace.

Market Competitive Assessments. In the area of market competitive assessments, an ongoing process needs to exists for analyzing the competition, oftentimes by the purchase of competing products for functional benchmarking and reverse engineering. Additionally, information about the competition's processes, and market strategies are gained through publications, trade shows and other marketing events. This competitive analysis is published for key members within the organization. Serious competitors are analyzed for development trends and the potential impact of their developments on your market segment. This information is oftentimes the key information that triggers a product launch. In addition to assessing the competition directly, new developments, technologies, and components should be researched for changes that can be utilized for competitive advantage.

Starting The Product. Once a determination is made to start a product, the first step is to formally establish the core development team and update the knowledge of these team members with regards to the market and its expectations. In order to do this effectively, specific ideas about the product or product features are defined by the team. This definition can be a written description, an illustration, or a partial model. The team utilizes the functional inputs from the three areas just identified: process technology, design enabling technologies, and market competitive assessments in developing the product ideas. These rough ideas are tested by several customers through a combination of customer sessions, telemarketing, and focus groups. The feedback and customer understanding should specifically identify the perceived value of the potential product, including pricing.

As the ideas converge and a competitive advantage emerges, then the concept design phase can begin. The concept design phase begins with the customer identified product specification. A QFD process can be used to define the full set of product features, and their relative importance to the customer. A strategic design plan can be developed which sketches out the customer usability of the product in its environment. The strategic design plan suggests alternative configurations depending on customer type and how the product may be used. It develops visualizations (renderings, illustrations, electronic models) around unique features with reasonable accuracy around size, geometry, human interface, and key technologies. From these documents, a new conceptual model can be derived, so that further customer testing can be done, and inputs from the design team and appropriate managers can be solicited.

As the new concept model is formed and the implementation technologies identified, the process plan can start to be developed. This plan identifies the technology risks, and suggests alternative processes and or design restrictions to keep the costs within the target range.

Developing The Detail. As the concept is converted into a functional specification, design engineering can begin detailed development of the product. The detailed development is done with manufacturing involvement so that the design is not only producible but utilizes the lowest cost processes. Throughout the design process significant use of Design For Manufacturing (DFM) methodologies are used to insure that the full impact of the design on the manufacturing processes is understood. Additionally, manufacturing engineers, and manufacturing suppliers provide recommendations based upon their experience with low cost processes. Manufacturing process are simulated whenever possible to predict the processing costs, and evaluate design alternatives. With regards to functional analysis, CAD created designs can be subjected to electronic design analysis, simulations, CAE tools, electronic prototyping, and other means to fully understand the design and to examine potential failure modes prior to its completion.

During this phase the design team is not only creating the design, but analyzing it to be predictable around the design functionality, performance, usability, manufacturing processes, cost, reliability, its ability to meet specific design standards, and the ability to pass design verification testing. Simulation is used as much as is practical, and other design characteristics are analyzed thoroughly--the goal being to correct any deficiencies while the design is still in the electronic format.

During this phase the market is continuing to be researched to find appropriate detailed information to feed it into the development process. Market sizing is developed along with the marketing strategies and specific marketing targets. The distribution strategy is being finalized with arrangements and unique configurations for the business partners to distribute. Pricing levels are being tested so that the functionality can be compared to the cost. Feedback is given to design regarding shifts in the market, including competitive positioning of the product, price/performance, customer feedback, and features being customer tested, as well as usability, appearance, etc. At the appropriate time, announcements, press releases, and customer invitees to see the new product are planned.

Testing The Product, Process And Market. As soon as prototypes can be built that reflect the final product, they are produced. A small quantity are built using a "rapid prototype" process to give engineering a head start on testing and to give marketing a chance to put a few early prototypes in front of customers for feedback. Manufacturing builds many of the prototypes on actual production lines in order to test the process prior to volume build. This provides manufacturing with the opportunity to discover potential bottlenecks that have not been anticipated, and to provide learning and training for the manufacturing personnel who will be producing the product in volume. The results of these steps usually means additional changes to the product.

Launching The Product. Product launch occurs when the design verification testing is successfully completed. In the concurrent engineering process this should be a "clean" event. The marketing plans are in place and being executed. The manufacturing personnel have been trained, and parts have been on order. Almost all design changes have been factored in at an early enough stage that any changes at this point are minor enhancements and can be phased in easily.

Collaboration Within The Process

When the concurrent engineering process is working well there are three distinct developments going on concurrently.

1. The concept design is evolving based upon market research
2. The development of the product
3. The development of the set of processes

All three developments feed one another. This collaboration is essential in the concurrent process. For example, the concept design is developed with the manufacturing processes in mind, such that the product is both producible and takes advantage of the most cost competitive processes available. The functional features of the concept design are based off preliminary market research, not the technical opinions of those defining the project. The concept is developed first as a specification, then into a physical model or series of models which can be market tested and reviewed by the various functions for their inputs. These inputs are factored into the next revisions of the concept design. Once again it is reviewed for manufacturing input, marketing input, and customer feedback and then the detail design can start.

The manufacturing process is constantly reviewed with cost reduction, production capability and producibility concerns. The idea being to use the most competitive, cost advantageous processes available. Oftentimes this requires understanding the minute details of each process step so that the production cost can be minimized through the elimination of partial steps, such as eliminating labor intensive masking steps prior to painting.

The concept design and market research is not complete once the detail design is started. The market research continues in order to more thoroughly understand the marketplace. Marketing also begins to develop the marketing messages and plans to advertise and distribute the product. This market research may turn up options or configurations of the product that may add substantially to the volume expectations or profitability of the product. Additionally this research may turn up other competitive advantages or disadvantages that must be considered. In the worst case, it may turn up information on market changes that require the cancellation of the product, or the addition of major options that will require months to factor in. But it is always better to find out market problems before the product is complete than to find them when the product is announced. The earlier the resolution or positioning of the product the better it will be received by the customer, and the easier it will be for the customer to recognize its unique advantages.

Because of the increased thoroughness of the development effort, the higher the likelihood of developing and retaining a competitive advantage, and the higher the likelihood of product success.

Setting The Goals And Objectives

In development, projects goals and objectives are fundamental in helping the functions and development teams focus on the correct set of issues. Therefore, goal setting by the senior staff is extremely important in order to establish the right environment for a successful outcome. Here are some guidelines:

1) Think Of Your Customer. What do your customers want, what functionality do they want, what price do the want, how do they want to be serviced, what would delight them in functionality. Part of the team objective may be to find this out in specific terms during the early part of the development cycle.

2) Position The Competition. Identify who the competition is by name and product. Identify the result you are looking for. Be specific as possible. Some examples are: to win against the competition in 5 major accounts when they are selling product X; to gain a foothold in the financial market segment; to retain your current customer base against very strong competitive pressure from product X.

Fundamentally Changing The Process 49

3) Identify Key Collaborative Connections. Identify the key new collaborative connections that need to be achieved with this development. Identify customers, suppliers, support organizations, consultants, and others who are needed to make the product a success.

4) Balance The Long Term With The Short Term. It is clear that progress with one product is limited in time, money, and resources. Therefore, it is important to structure achievable goals in the immediate product teams, and subsequent teams can build upon the learning and achievements. This linking from project to project is a key role of the senior management team. Senior management is often criticized for its short term month to month mindset, but by establishing objectives that can be built upon from one team program to the next, the long term improvement objectives can be recognized and achieved.

5) Product Goals. Using these simple rules, product goals can be set at the correct level. Goals at the functional detail level have the potential to trap product teams with the inability to find the right solution for their market. For instance, use of a specific portion of a previous design may in reality be too restrictive, and limit the product to the existing customer base. This decision is acceptable if it is based upon research on the market segment the product is targeting.

Product Development As A Structured Process

Structured development processes have been in existence for many years within larger companies. They have been put in place to provide a system for balancing control and the work of the development teams. A typical development process has several phases. At the end of each phase there is a criteria for completion of that phase and a mandatory management review for proceeding to the next phase of the program.

Typical Phased Development Process. The typical phased development process has several steps. In a more serial process they may look something like this:

Phases of Development
1. Strategy and Requirements
2. Planning and Preliminary Design
3. Design Development
4. Design Verification
5. Manufacturing Start-up

These processes were put in place to establish a quality level in the design process--the theory being that a thorough process, although painful to the engineers who had to endure it, would result in an improved product, and one that the company was ready to offer to its customers. Thus, the reviews became the thrust of the process, and readiness for these reviews became the task of the team members. Successfully passing the reviews meant that they had done the correct development tasks for the money allocated to their projects. Thus, it was these review processes that took on a life of their own, and deviations from the process became a cause for great alarm. These companies did not realize that these older well-established development processes were the exact thing that prevented them from evolving into a more progressive concurrent engineering environment.

Concurrent Engineering Phased Development Process. In a more modern concurrent development process, the emphasis is on collaboration, process, and technology, not steps. At each measurement review along the way, five topics are reviewed.

1. Market research and preparedness

2. Design concept, development and schedule
3. Manufacturing competitive contribution and readiness
4. Collaborative partners' contribution and risk
5. Technology decisions, contribution and risk

The phased steps for review are as follows:

1) Starting The Product. This is a review of the process technology selections, the design enabling technologies to be used (including patented competitive advantages), the original market competitive assessments used to launch the product effort, and the information technology selected. Additionally, it is a review of the recent market and competitive research done by the development team, their concept specification, their identification of product competitive advantage, and manufacturing's technology risks and supply side risks. At this point, a preliminary product specification should exist based upon initial customer inputs (but be allowed to change as the team learns more about the market), and manufacturing cost assessment should be identified.

2) Developing The Detail. This is a formal review of a product that has been developed to a detailed level. A reasonable amount of simulation has been completed to understand that the product will function correctly. Models have been built for design evaluation, market evaluation, and manufacturing evaluation. Manufacturing has provided a reasonably accurate cost estimate (+/- 10%), with identified opportunities for further cost reductions. Design analysis, electronic technology CAD and CAE tools and laboratory testing are all available for review. Market strategies are presented, along with an update on the competitive and market research. The reasons certain concepts were chosen over others is identified. Competitive advantages are defined in detail, and any collaborative partners identify their unique contributions and risks.

3) Testing The Product, Process And Market. This review is done when prototypes have been built--ones that reasonably reflect the final product. Marketing has had a chance to test the product ideas with selected customers, design has functionally tested the product to ensure its proper operation and manufacturing has built small numbers of the product discovering any bottlenecks, producibility issues, etc.

4) Launching The Product. This is the final review, and is done when all the design verification testing is complete, manufacturing is ready for volume production and marketing is ready for product announcement and launch.

Managing a project in this way will allow the collaborative efforts of the teams to be the primary focal point for product development. It will also allow management sufficient review of the progress to spot any significant risks being taken by the team. At each of these reviews except the last one, there should be the expectation of change. For instance, the development team might be refocused to put more emphasis on the differentiating features. The development schedule may be accelerated by adding more engineers to the development process. Conflicts might occur among organizations over standards implementation, requests for additional funds, prioritization of tasks over schedule or vice versa. These are only some of the issues senior managers need to resolve in order to keep the development team properly focused.

The Effect Of Development Time On Profitability

Shorter development cycles win market share. Because of this phenomenon, it is important to understand the relationship of development expenditures and market share (product volumes) to product profit. These relationships can be seen in Figure 10. In the first curve the income from the development effort is shown by

Fundamentally Changing The Process 53

the solid line. At first it is negative since the development is occurring without any revenue being generated. It turns positive

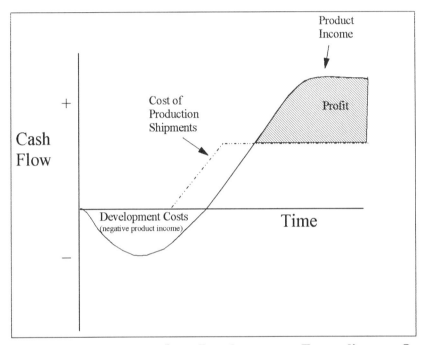

Figure 10 Return On Development Expenditures--On Schedule. *New product cash flow for an on-schedule product under development. Note that product profit (shaded area) needs to exceed the development costs (negative area) in order to give a positive return on the development expenditure. This return is dependent on the longevity of the product in the marketplace and the volume of product reached.* (Courtesy Prof. Dan Shunk Ph.D. Arizona State University)

only after production units start to ship, which is shown by the dash line, and the development costs decline. The profit is the difference between the income and the cost of production, assuming that the cost of sales, marketing and administration is included in the cost of production. Total profit is the area between the two curves. If the product life is only as long as is shown on this graph, then the profit only marginally exceeds the cost of development.

The dependency of the return on the new product expenditure is dependent on many factors, including volume, price, and time in the marketplace. All of these are improved with shorter new product development cycles.

In the next example shown in the second chart, Figure 11, the start of production shipments has slipped as a result of design delays. This has four effects. First, the development costs increase due to the extended time. Second, the volume of units is not as great since market share is affected by lateness. Third, the net profit is substantially reduced due to increased price competition. Fourth, the return on investment for the development is negative, since the profit does not exceed the initial investment. Lateness is disastrous to the financial metrics of product development.

Rapid Product Development

One of the fundamental reasons for converting a process to a concurrent engineering process is to achieve a substantial reduction in development time. It should be noted that rapid product development takes the idea of concurrent engineering to the next level of development cycle reduction. Rapid product development requires both concurrent development and time compression of each of the major steps. The time compression happens due to the following reasons:

>Efficient and effective collaboration
>Automation/Information technology
>Functional preparedness prior to starting
>Structured methodologies used during development
>Effective project management

Fundamentally Changing The Process

Avoiding management stoppages
Skilled individuals/ team members
Adequate resources

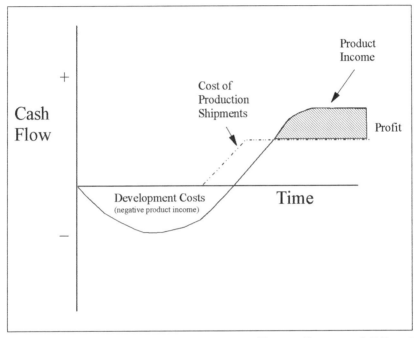

Figure 11 Return On Development Expenditures--6 Month Slip. *Note that the product profit does not exceed the development costs. In the case of a product slip, this negative return is not understood when the development expenditures are made. In most cases these products will continue to be completed since the incremental cost to complete is substantially less than the profit expected.* (Courtesy Prof. Dan Shunk Ph.D. Arizona State University)

Efficient And Effective Collaboration. Collaboration is one of the most important elements of concurrent engineering. It requires both efficiency and effectiveness. It is important that anyone participating or contributing to the development effort does so by providing useful information in a timely manner. Many of the delays in a development process consist of waiting for key information from a team member or contributing source. For instance, the price of a tool may cause a change in the design to avoid the cost. The decision cannot be made until the information is provided from either manufacturing or the supplier. Collaboration requires the utmost in cooperation, since very few design decisions come from simple answers, but rather an exchange of information about specific details of a process, or part or requirements.

Automation/Information Technology. Automation has historically been a major method for reducing cycle time. It is also true in the development environment. The faster and easier the tools are to use, the faster the process. The information technology tools for the development process provide the means for improving the overall development cycle, the quality of the design, and the cost of the design. Graphical users interfaces are emerging as the primary method of computer usage; thus, "point and click" with a mouse replaces complex command language. Electronic communications provides the means to capture ideas, and share them in real time. Document management tools, and product data management tools allow the product development data to be shared and controlled. Workflow technology allows the rapid development of concepts into designs and the process to be time management. Automated project management tools allow the team to see the impact on the total project of one team member's deliverables or critical decisions. Computer-aided engineering allows simulation of the electronically developed product before it is committed to physical representations. Through simulation and analysis, the engineering team has a good idea of the product's ability to function correctly before submitting it to design verification testing. Thus, information technology used in the development process provides the fundamental infrastructure for speeding up the complete development process.

Fundamentally Changing The Process

Functional Preparedness Prior To Starting. Another key to rapid development is functional preparedness prior to starting a project. For example, if the product team can simply pick technologies that are being developed by the function and have an expertise within the function to rely upon as a consultant to the development, then the development process can proceed rapidly. If, however, the team has to do its own technology research and develop its own expertise, the cycle will be greatly extended. This may even put the product development at risk. Another example is marketing preparedness. If competitive research is an ongoing activity, the development team can rely on their inputs, rather than developing their own competitive data. The examples go on and on. The more the functions can do independent of the development teams, the more rapidly a product can be developed.

Structured Methodologies Used During Development. Structured methodologies provide a way for various opinions to be discussed in a constructive manner, and put into the context of the development effort. These methodologies allow inputs from various parts of the operation to have a meaningful impact on the design itself. For example, Quality Function Deployment (QFD) allows customer inputs, competitive inputs and price to be compared in a way that extracts the key design features on which to base the design. Ad hoc methods such as working meeting discussions are more subjective and less structured, and often lead to much debate and delays in the development process. Providing structured methodologies for the development effort allows the discussion to take place in a useful framework and the decisions to be made and understood by the team and other collaborating groups.

Effective Project Management. Product development activities that involve multiple players, functions or collaborative efforts, cannot happen in a timely manner without effective project management. Many people would rate this as the number one item for improving design cycle time. Project management requires leadership in decision making, problem resolution, creation of a collaborative atmosphere, meeting management, project schedules, cost control, task assignments, role clarity, and management interface to name a few of the key tasks. An effective leader can overcome many barriers and hold team members accountable to meet tight schedules and constrained budgets.

Avoiding Management Stoppages. Many companies do not appreciate the impact that their management processes have on the development teams. There are many examples of this problem. One simple example is the policy that prevents engineers from talking to suppliers. Engineers need manufacturing input on the part being designed. The policy often implemented by purchasing departments is intended to avoid confusion at the supplier, and the potential for excessive prices. Yet without the suppliers input, the design may end up much more costly. Another example is the review process used by companies. Many companies treat management review steps as "gates" to further developments, thus effectively putting the development on hold until the management has approved the design in its current state. These polices designed to provide effective controls within the organization oftentimes cause considerable delays for the development effort.

The reviews themselves are costly. Even though none of the reviews are very long, usually one to five days, the summation total of time is very significant. Over the course of the development cycle the time spent preparing for reviews, attending reviews, chasing information to prepare senior managers, all adds up to a noticeable percentage of the complete development cycle. Some estimate the time spent on reviews, management indecision, and cumbersome bureaucratic processes to be as high as 35% of the complete development cycle.

Engineers caught in these situations need to pay special attention to the implications of management delays. Some

Fundamentally Changing The Process 59

techniques for presenting key choices for decisions can be used (such as the Pugh Process discussed in Chapter 4). It can clarify the facts and the decision process. In addition, obtaining the support and attention of key management members who can keep these decisions in the forefront is important. During delays and political processes, the squeaky wheel supported by sound technical data will get the grease. Where cumbersome polices exist, such as in communication to vendors, one should work towards exceptions to the policies in key areas. Exceptions are normally easier to obtain, whereas sweeping policy change is time consuming, lengthy, and represents a higher risk of failure. Lastly, if the decision process seems inordinately slow, request that management define the steps required before a decision can be reached.

Engineering managers faced with this problem need to note the exceptions to the normal management processes and change the management processes to accommodate the speed and collaboration of the CE environment. In one recent CE implementation, it took the vendor two days to produce a rapid prototype part and deliver it. It took the company three weeks to issue the purchase order to the vendor.

Skilled Individuals/Team Members. Training and knowledge are key ingredients in rapid development. Those individuals who are knowledgeable about the product they are working on, the market they serve, and the manufacturing process they will use, and who are collaborative by their very nature will develop the most cost competitive designs. Thus, broad skills are very important for effective teams. At the technology level however, very specific skills are required to give a competitive advantage. Focus is the key. Thus, even though senior team members have broad skills, technology contributors have very narrow skills with significant depth.

Adequate Resources. The amount of resources in both people and dollars is difficult to determine without previous history in the development effort. Previous expenses for a successful product should be the guiding factor. Many a design team becomes too constrained through limiting their resources. Engineers often overestimate their own abilities, and underestimate the dependency on others to achieve the task, and as a result slip schedules because they have not completed their tasks on time. Given a choice it is much better to have many experts on a part-time basis than to have a few individuals who need to learn prior to contributing. The learning through doing phase represents high risk. Providing adequate skilled resources will prevent the delays associated with overloading and resource justification. Providing skilled support for those resources that must be new will provide a way of limiting the risk of first time members on a new development effort.

The Three T's Of CE

Many items regarding the implementation of concurrent engineering have been covered in this chapter. To refocus the thinking, it is important to remember the three key elements of concurrent engineering:

>Collaboration
>Process
>Information technology

A simple way of focusing the initial steps in implementing concurrent engineering is to remember the basic underlying ideas that are the foundation of the above three items. These are referred to as the three T's of CE. They are:

>Teams

Fundamentally Changing The Process 61

Tools
Techniques

As these are implemented, a major piece of the conversion to concurrent engineering will take place. A chapter in this book is devoted to each of the three T's.

Summary

This chapter provides an overview of the changes needed to convert an existing process into a concurrent engineering process. The fundamental changes are highlighted. Where to start is suggested. Specific shifts in the development spending are shown. The development process using concurrent engineering was defined at an upper level, as well as in specific areas the supporting functions need to develop in order to have an effective concurrent engineering environment. Specific emphasis was discussed concerning collaboration within the development team. The four basic steps for product goal setting were established. The management review process as a phased development review process was contrasted to the historical process. Specific areas that are needed to achieve a rapid product development, by not only operating in a parallel fashion, but also by reducing the time it takes to complete each step, were defined. Specific diagrams and charts were shown identifying the development process flows.

Chapter 3
How To Form A Concurrent Engineering Team

The most fundamental component of concurrent engineering is the idea of teams. Simply stated, this is the participation of concerned departments as early as possible in the design cycle in order to maximize their contribution to the product and to its success in the marketplace. This participation is often regarded as more important than most other aspects of the process. Forming a successful team is not easy.

When forming teams one must ask the question, do I have a team or simply a collection of people. The notion of team is a very distinctive notion. For instance a manager's staff may not be a team or act like a team. A team has specific characteristics that vary depending on the type of team. This chapter discusses industrial teams and how to effectively utilize this organizational concept in the concurrent engineering process.

What Is A New Product Industrial Team?

There are many types of teams in today's culture. There are basketball teams, football teams, bowling teams, baseball teams, parent-teacher teams, quality circle teams, project teams, product development teams, and many other teams. Even several horses pulling a wagon are referred to as a team. What makes up a team? *Webster's New World Dictionary* defines teams as "A group of people working together in a coordinated effort". It defines teamwork as: "A joint action by a group of people, in which individual interests are subordinated to group unity and efficiency; coordinated effort, as of an athletic team". This definition only partially covers the complexities of a new product industrial team.

Industrial teams require the contribution of many individuals to achieve the outcome. Many of the contributors are not team members even though they supply critical pieces of information needed for the team's success. Some examples are purchasing or component engineering. Thus, unlike sports teams, the industrial new product team is not a self-contained group.

Additionally, many of the decisions a team makes are dependent on funding and resources allocated to the team's use. This is unlike sports teams, whose play calling is not dependent on resources allocated, since the number of people on the playing field is constant.

Furthermore, technology and market expectations are constantly changing. These fundamental changes affect the basic underlying values in the marketplace. Sports teams do not have this uncertainty to deal with, because the rules for them are the same. The uncertainty for them is only in the method and skills used to interpret and implement the rules.

Thus, industrial teams are more different in makeup and operation than is commonly understood by the cultural concept of teams. This chapter will shed some light on product development teams, including the roles of the team members, team structure, leadership, team processes, etc.

Contrasting Industrial New Product Teams With Sports Teams

A simple way to understand the differences in team concepts between industrial teams and our cultural understanding of sports teams is to compare the two.

Sports teams have clearly defined rules and referees. Industrial teams seldom have any rules. If a team member is treated unfairly or is blamed erroneously for an error, seldom is the full truth recognized. Penalties are rarely assessed.

Sports teams are self-managed. The coach calls the plays and all the dependencies for carrying out the play are within the team's control. The sports team is self-contained. The industrial team may claim to be self-managed but it cannot act independent to the direction of management, nor is it possible to contain all the functions, suppliers, and technical experts upon which they are dependent and still be a functioning team.

Methods for sports teams are clearly defined. The moves and techniques are practiced and strengthened. Often weight training and special diets are used to hone the athletes abilities. For industrial teams, the specific methods are usually not known by the team before they start. Each team operates differently. The methods are selected and learned during the development. If they have been used before by one or more individuals on the team, that is viewed as an adequate understanding for the new implementation. Product development methodologies are not normally practiced and honed outside the development process.

Other areas to compare include how the competition is studied and how plans are developed to overcome the competition's strengths and attack their weakness. The comparison can go on and on. A brief summary is shown in Figure 12.

Category	Athletic Team	Industrial Team
Clear Rules	Rule Book	Undefined
Fair Play	Penalties Referees	Seldom called Never watched
Ability to Self-Manage	Self-Contained	Dependent on others
Team Goal/ Measurement/ Timeliness	Win/ Points/ Immediately known	Win/ Sales / Long Term
Individual Measurement	Statistics/ Recognition	Unclear/ Perception Cooperation
Leader	Coach	Team Member
Roles	Defined/ Practiced/ Extensive training	Assigned/ Unclear/ Little training
Methods	Practiced/ Signals for coordination	Selected/ Ad Hoc coordination
Competition	Scouted/ Filmed/ Planned attack	Opinions/ Trade information
Structure	Positions understood/ Player recruited/ Best in class	Positions not clear/ Use available players

Figure 12 Athletic Teams Versus Industrial New Product Teams.

Figure 12 contrasts some key structural differences between an athletic team and a typical industrial new product team. Fundamentally, industrial teams are not measured, structured, supported, or defined as are athletic teams. Also, their success criteria are not as clear or short-term, nor are the leadership, direction, and feedback as well defined, as they usually are for sports teams.

Typical Teams In Industry

Teams typically are comprised of members of several functions including engineering, manufacturing, marketing, and quality to name a few. These individuals are pulled together with various guidelines, rules, and constraints regarding their activity. These guidelines, rules, and constraints form the team concept used by a specific company. Many types of team concepts are used in industry today. Here are two common ones and why they don't work.

"Over-The-Wall" Syndrome. Some teams are constructed with the functional organization simply assigning members of their staff to the development team. In this approach the functional member is still measured by the performance of the function. Thus, those items that disrupt the functional measurements are viewed as negative. Typically, in this model, engineering is measured on design completion and development cost, and manufacturing is measured on output, manufacturing cost control, and manufacturing lead time.

New product introductions into manufacturing under this structure are very difficult. Engineers are anxious to complete designs to make their measurement. Manufacturing does not want to spend any money not allocated, especially for items such as late design changes. Nor do they want to disrupt the manufacturing production line, even to learn the new product. It becomes the manufacturing team member's role to protect the manufacturing measurement. Thus, a wall emerges, and manufacturing won't "sign-off" on a design and accept responsibility until they are absolutely sure every detail is worked out and all costs are accounted for. Engineering, on the other hand, wants to complete the task, and can't stand the manufacturing "pickyness" as they perceive it. They worry about development dollars being spent on these last minute corrections, (unwisely in their opinion) as well as committed schedules being missed.

In this structure the behavior is supported by the management metrics. Since the individuals and their managers get measured on the metrics, their decisions and actions often have career impacting consequences. Thus, the attitude of passing it "over-the-wall" emerges. Manufacturing views it as being stuck with the design and its flaws. Engineering wants to walk away and work on more creative tasks. In this measurement system there are no winners. Since both design and manufacturing engineers are imperfect people some non-optimum design features make it into manufacturing.

Matrix Management Teams. In the matrix management approach the functions plan for new product introductions by assigning a functional representative to the team. The representative from engineering is usually chartered to develop the project, and the representative from manufacturing is representing the various subfunctions within manufacturing. Marketing represents marketing and sales. This method is suited for bringing information about new designs to the functions, but is poorly suited for the functions' ability to influence design. In this method the team member brings subfunctional contributors (for example manufacturing quality) to influence design engineering, and then the team member (manufacturing) must sustain the inputs for the remainder of the design. When it comes to decision making the team member is caught in a no-win situation. Product compromises will result in a loss of support from the team member's functional group, or alternatively a resistive approach towards design will subvert the team support.

A representative approach rarely works over time without being very costly since it doesn't take very long before subfunctions begin to want their own representative at the team meeting. For instance, in manufacturing, if the main functional representative came from manufacturing engineering, the materials organization would want a representative, then quality, etc. This approach leads to a very complex set of teams, and a very costly development effort.

In addition, the team members who represent the functions are required to go back to their constituents, and get approval and acceptance of their plans. This in itself is a very time-consuming task since many decisions get revisited in the process. New product development needs to occur at a very rapid pace, oftentimes with decisions being made at the team meetings. A representative process subverts the team's independence and flexibility, and draws too many decision makers into the process. An old saying in cooking applies here: "Many cooks does not a better broth make".

A Concurrent Engineering Team. A concurrent engineering team is very different. The team is chartered with specific tasks. The members are all contributing to achieve those tasks. They are not functional representatives, even though they come from the functional organization. Nor are they measured by the same measurements of the function. They have separate new product development measurements. Their goals are based on the product and market and typically are as follows

        Shipment date
        Market penetration goal
        Revenue goal
        Product cost goal
        Development cost goal

This team is very collaborative in its nature, with each member using his expertise and resources to further the goals and objectives of the team. See Figure 13.

The goals are clearly directed by one person, either the director of engineering or the director of marketing. There is

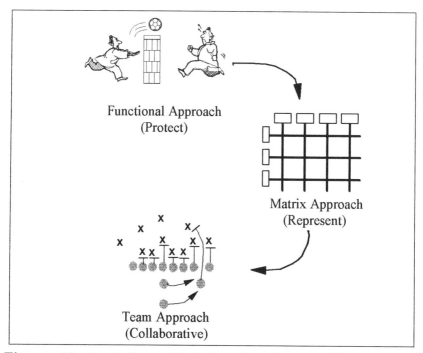

Figure 13 Evolution Of Industrial Teams. *New product introduction teams have been established with different objectives. The functional approach is dedicated to improving detailed functional metrics which often conflicts with product objectives. The matrix approach is dedicated to working coordination between the functions. The team approach is collaborative and dedicated to the success of the team and product.*

(Some of the artwork is derived from Lotus SmartPics for Windows. ©1991 Lotus Development Corporation. Lotus and SmartPics are registered trademarks of Lotus Development Corporation)

support to guide and coach the individuals in their roles. Their roles are significantly different from non-new product positions. For instance, the manufacturing engineer's role is measured on the cost effectiveness of the design when compared to the external competition. This requires a great deal of knowledge that does not originate form the normal manufacturing engineering function.

Aim Of The Collaborative Team Approach

The aim of the collaborative team approach within concurrent engineering is:

1. To intensify cooperation.
2. To establish an atmosphere that fosters early and effective communications.
3. To allow for early unfinished releases of the design to be reviewed by the team members for their contribution. This review is done without formal departmental procedures interfering in the process.
4. To make the functional team members responsible to check for possible product and business risks, avoiding downstream pitfalls.

The recent involvement of suppliers on the teams is intended to provide supplier input on manufacturability and part schedules. The recent involvement of customers is intended to provide more customer awareness within the design team.

Concurrent Engineering Requires Early Involvement

The concurrent engineering team approach has been very effective in breaking down the wall that existed between design and support groups. Design engineering has often been reluctant to release its designs until the engineering was near completion. In mechanical design this often meant several steps were completed before drawing release. The geometry was defined; the designs were technically analyzed for design characteristics such as structural and tolerance analysis; there existed a high degree of confidence in the proper functioning of the design; and the design engineers were confident that the product could be manufactured in some sub-optimal fashion.

Manufacturing and marketing on the other hand have significant logistical issues. Many departments have long lead times and involve large numbers of people in the implementation. These organizations want to get information early on so that they can begin their processes in both development and communications. Early information for them allows the communication network to become established, allows for partial early development, and allows for the logistics to begin being planned.

With the team concept, the idea is that the support team members should be able to gauge and manage the risk of early release of the design data. In addition, they contribute to the design. The support team members need to provide input that is specific, technically accurate and represents the best ideas, since this information will be reflected in the final product and the final business results.

Establishing CE Teams

A CE team should have a clear written charter. This defines the team member's authority and role and prevents them from falling back into the norm of representing or protecting their function. The intention of the charter is to avoid misunderstanding or confusion. Some of the items in the charter include:

1. Brief outline of the product

2. The competitive situation
3. The purpose of the team
4. The goals of the team
5. What the team has the authority to do (and not to do)
6. The role of each member
7. The level of problem solving, decision making, implementation requirements, budget authority
8. The senior manager accountable for the team
9. The team membership, and key informational reviews with non-team functional groups
10. Support and commitment from the CEO or appropriate vice-president

Assigning Team Members

In the concurrent engineering process it is important to assign the full set of core team players at the beginning of the process. This allows them to leverage their expertise and the expertise of their function early in the development process. Historically this has not been the case. In the standard phase review process for development, team members where assigned at different phases of the process. See Figure 14.

The Role Of The Senior Manager

This person acts as the liaison and can speak for all of the functions. This is necessary so that quick answers can be obtained. Since committees will have conflict, and all opinions will not prevail, the team needs one person with whom they can turn for assistance. This person is also the person who approves the product, approves the development plan, and who provides the pass or fail decision at the management reviews. This, however, is not the team leader.

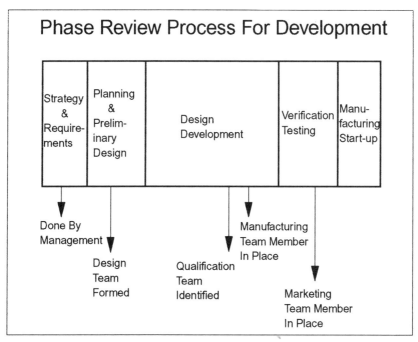

Figure 14 Sequential Process - Team Assigned. *In the typical phased development process, many of the key team members were added too late in the development cycle to have any effect on the design itself. They were added in what was believed to be a "just-in-time" approach, just before they were needed to execute the design roll-out. In contrast to this approach, the core concurrent engineering team members start at the very beginning of the strategy and requirements phase.*

Concurrent Engineering Team Leader

The team leader is usually a member of the engineering department. This person chairs the meeting, manages the budget, and the project schedule, calls the team meeting, and holds the review meetings. This is management's main contact with the product team. This person is chartered with resolving conflict, recruiting new members, developing the plans, organizing the team, and coordinating the activities. The team leader, however, is not the boss of the team members. The leader does not have the sole responsibility for managing and facilitating all of the tasks. Each team member should play a part in both management and facilitation. Each member contributes to the plan, and is responsible for organizing their own contribution and the contribution of their function. When major issues arise, the appropriate team member takes the lead in solving the issue or preparing the information for management review.

Team Participation

For teams to be effective, each team member must contribute to the success of the team. Full participation should be encouraged. Each team member should be looking for ways to keep the group moving forward. A major document in providing this coordination is the team's plan. This allows the team to clarify its intent and solicit the support of fellow team members. It captures the team's ideas for meeting the challenges ahead. It also provides a method of measuring progress. The plan should include the steps to be taken throughout the development process. This can be done in a flow diagram or in a milestone chart. It is important to identify timing in the plan so the tasks and expenses can be adequately controlled. The plan is also the vehicle for management approval.

The Team Plan

A team plan should include the major elements of the activity. Here are the categories that should be used at the very beginning of the development cycle:

Product Plan by the Team
  Team charter
  List team members
  Market requirements/changes
  Key competing products
  Customer visits/information collection process
  Projection of customer needs
  Projection of competitors market positioning
  Identification of key technologies to be used
  Identification of key competitive advantages
  Identify manufacturing impact/risks
  Identify applicable standards
  Identify applicable regulatory requirements
  Develop preliminary schedule
  Identify development budget requirements
  Preliminary business projections
  Competitive market position
  Volumes anticipated
  Projected product cost/price
  Product distribution requirements

Design Engineering's Role On The Concurrent Engineering Team

Design engineering's role as a team member on the concurrent engineering team is to provide the design activities. These include concept design and development, product design, design analysis, simulation, design testing and design documentation. Some things do change for design engineering when compared to a sequential development process. For instance, the team recommends the final concept to be developed, and the senior manager approves or disapproves of it. The team members also approve the final version of the detailed design. Thus, it is important for the design engineer who is on the core team to clearly understand the inputs from marketing, manufacturing and other team members and make sure their inputs are included in both the concept and in the detailed design. On the design engineers part, this takes a willingness for collaboration, a willingness to change first and second pass creations to meet the needs of the team and to factor in the feedback from early customer reviews. The engineering team member has to focus on success in the marketplace, not success in the world of design. Oftentimes this means that the latest technology may not be used, or the latest features may not be incorporated into the design in order to meet the customer's cost objectives. Designs are set to match the market expectations both current and future. How closely they match these expectations means the eventual success or failure of the product.

Team's Role In Setting Future Market Requirements

Setting the future market expectations results from a combination of activities, including customer visits, market research and competitive analyses. It is important to include all the major contributors to the design in these analyses to take advantage of their insight and for their own education. Where it is affordable, a competitor's product should be purchased, functionally benchmarked, and then torn down to examine its design. For companies that are behind in the design activity, this will energize the design department into trying competitive approaches that they might have been unwilling to risk earlier. It helps the engineers to realize how competitive their skills are or more importantly how competitive they aren't. Manufacturing is a key player in the competitive analysis activities as are critical part suppliers. They can point out processes that may be in use by a competitor that are more cost competitive than the ones that are the basis for their current product set.

Concurrent Engineering Teams

Competitive Teardown Importance

The importance of multiple team members and their support functions looking at the competitive products should not be under-emphasized. During one such analysis performed at a major company, the design engineering team reviewed the product and concluded that nothing unique existed in the competitor's design, except the final assembly snap fits. They estimated the cost savings to be minimal and not worth the cost to implement. The manufacturing and marketing individuals looked at the snap fits, and saw a creative way to manage the complex configuration problem. Their own product required a significant number of options to be integrated and configured at final assembly adding four to six weeks into the lead time for products. With the competitor's design, much of the final configuration could happen at the resellers. In this design the reseller would order a base configuration and options from stock and simply snap them together prior to delivering it to the customer, reducing the delivery time by several weeks.

Managing product configurations in this manner allowed the competitor to gain the cost advantages of high volume manufacturing by mass producing base line configurations. It also allowed the resellers to provide a value added service for the end customer. Thus, the team better understood their competitors' product distribution strategy, and why resellers were so loyal to the competition, i.e., the resellers' value was designed into the product.

Marketing's Role

Marketing has a major role in the concurrent engineering team. Its role is to understand the market basis for the product, the customers' needs, the competitive advantages, and the positioning of the product in the marketplace. It is also responsible for the announcement activities, the advertising activities, the distribution methods for the product, developing the relationship with resellers, and providing the messages to the marketplace that will help keep the product selling. They set up the customer surveys, provide the market analysis activities, and provide the final market requirements document to the team.

The final market requirements document is a combination of the inputs from all sources. It becomes the focus for the development effort, and a referenced activity that helps to keep critique focused, and to avoid the "ever changing specification syndrome". Done correctly, it will become a learning activity for the senior management staff. It will identify the key features of the product, and provide priority to them. It will identify which features are critical to the customer and which ones are convenience items. Thus, the appropriate effort can be placed on the critical features by the design team, making the critical features the ones to use in providing competitive advantages.

Because of the importance of matching the requirements to the market, in some industries the marketing manger may be the team leader.

Manufacturing Team Member

The manufacturing team member has a critical function to play in the development of the new product. The manufacturing team member identifies the processes that will be used for producing the product. The manufacturing team member should be aware of the latest technology developments and process changes that are occurring within the industry. This person identifies manufacturing's key technologies that can give the product a competitive advantage. Some examples of these might be fine pitch high gate count ASIC devices, double-sided surface mount, emerging technologies in flat panel displays, etc.

The processes recommended need not necessarily be the latest in technology, but alternatively could be smart and cost effective uses of existing processes. Some examples are: designing the product to minimize the number of masking steps prior to painting; or developing plastic parts that do not require slide action tooling; or minimizing multi-step processing through creative design.

Simulation tools exist for many manufacturing processes. These can help the manufacturing engineer understand potential manufacturing problems and provide this feedback to the design department at an early design stage.

Additionally, the manufacturing person develops alliances with key suppliers, and utilizes them to minimize costs for many items. The suppliers also communicate future process trends and expected cost containment opportunities. The manufacturing person identifies any new processes and manages the risk for the development team. They identify how the new processes will be qualified for volume usage, and what the alternatives are to the proposed process solutions. The manufacturing person identifies the sole source parts, and possible alternative parts in cases of production and supply issues. In some cases, the manufacturing person identifies the risk of part obsolescence affecting both the prices and the company's ability to deliver products to customers.

Oftentimes the new product requires new equipment, fixtures, and tooling. The manufacturing engineer specifies the new equipment to be purchased, identifies the implementation scheduled for the new process, and identifies the qualification steps required to insure consistent product quality.

With regards to prototypes, the manufacturing person identifies how prototypes will be developed, out of which materials, and the expected turnaround time and cost of each. He identifies the lead time required to begin the first prototype builds in production, and works with the rest of manufacturing to make this happen rapidly and effectively.

Remaining Team Members

Depending on the team and the scope of the effort, more people can be added to the development team directly. These roles should not conflict with any team members currently on the team. For example, a person responsible for testing the product or a person associated with maintaining quality in manufacturing may often be added without confusion.

If people are added after the start of the product, the team risks some loss of time bringing the new individual up to speed, and in listening to the new ideas the person brings. Oftentimes this is a frustrating experience at first, but usually results in keen new insights. A typical team might contain the following people:

> Engineering team leader
> Lead engineering manager
> Marketing manager
> Manufacturing engineer
> Service engineer
> Information technology specialist
> Other key engineering team members
>> Software engineer
>> Mechanical engineer
>> Qualification engineer
>> Technology specialist
> Concurrent engineering mentor (situational)

With information technology playing a key part in the effort, the team should include an information technology specialist. This person can make sure the right technology tools are appropriately used and that people are appropriately trained. If any problems arise that will cause delays due to the information technology, this person is responsible for rapid resolution to keep the team on schedule.

CE Mentor Role

The CE mentor role is situational and is an important one for those companies just starting out in a concurrent engineering environment. This person is the one who has previous experience implementing CE and can help the team to convert their old style of thinking into the new methods of CE. This person works with the team and one-on-one with team members and managers to help teams and individuals think about the CE environment and how it's designed to operate. This person also trains team members in techniques, and elevates systemic problems that inhibit the CE process. This person documents lessons learned and passes them on to other new product teams that may or may not have a mentor. The primary advantage of a mentor is that he gets the team fully into the CE mindset sooner and quicker. With some companies who's ingrained methods will be the most difficult thing to change, this is the preferred method to convert teams.

Organizational Support For Effective Concurrent Engineering

Organizations need to support the collaborative environment for an effective concurrent engineering implementation.

Marketing Functional Support

The marketing function should have available several capabilities including a market analysis capability that allows market positioning, and the determination of new directions for product development. The supporting fundamentals for establishing this capability are access to a market research library, a competitive database, market research reports, market partitioning data, customer support forums, and partner relationships. This function should also help the marketing team member target the product to specific markets. This function helps to specify and target customers and partners in this market to provide feedback.

If a sound foundation exists within the marketing function, this will form the basis of very competitive products. If a weak foundation exists, then the marketing person needs to be aware of the potential sources of information and where to acquire the information as a part of the development process. With a weak foundation, the risk of an inadequate understanding of the market, pricing, distribution, and customer expectations is very high.

A strategic process that should exist within marketing is competitive benchmarking and analysis. This analysis drives key development decisions by senior managers. In addition, the standard market functions of product information, advertising, direct mail campaigns, lead generation activities, press releases, etc., should exist, and be called upon to support product announcements and start-ups.

Manufacturing Functional Support

In manufacturing, the functional collaboration and communication effort needs to be paramount. Manufacturing by its nature is a grouping of specialists who work together following specific and detailed plans. New products, because of their start-up inconsistencies and rapid product changes, create problems for those manufacturing plans and activities. Communication and collaboration is the key in manufacturing to allow new product flexibility while minimizing the disruption to existing plans. In manufacturing, an infrastructure should exist that helps the manufacturing team member to communicate the requirements to the various subfunctions, to coordinate their new product activities, and to receive appropriate information to feed back to the new product team.

A research infrastructure needs to exist within manufacturing that provides knowledge about cost competitive manufacturing processes. This research activity includes internal research, literature research, key suppliers, and strategic relationships. The process development activity is another separate manufacturing development activity that provides new process capabilities that result in product competitive advantages.

Manufacturing research and process development is a very important function to the CE concept. Tying these critical development efforts into new product development is a fundamental part of the manufacturing function. It normally requires a manufacturing engineer who establishes the process rules for product design, educates design, and handholds the first few implementations. Additionally, the manufacturing function needs to provide access to the suppliers' new processes and products for the design team. This is vital for processes not developed by the company, so that products can incorporate the latest technological advantages. For small companies, this is the predominant method of obtaining technological competitiveness.

The manufacturing function should have an easy method to rapidly produce prototypes, and have a materials planning function that works closely with the development effort to factor in the constant change in plans, parts lists, suppliers, etc. to help keep the project on schedule.

The manufacturing function requires a competitive analysis function. This function complements the competitive analysis function in marketing. This manufacturing function costs out the competitive products and provides insight into the manufacturing processes that are in use by the competition. This function provides critical cost information to the development of the products. This function is especially important if the company is behind the competition on the technology curve, since it will highlight the cost disadvantages of being behind the technology leader.

Engineering Functional Support

The engineering function provides team member support. The support activities include product research, engineering standards support, laboratory support, information technology support including applications, and technical consultants and specialists to help with difficult areas. The support function allows team members access to ongoing activities so that their product development can take advantage of the latest technologies, product research and software applications. In addition, the engineering support function provides the ongoing laboratory facilities that engineers can use to test out new ideas and concepts prior to placing them into the product design. This function often includes the design verification, reliability and qualification personnel and facilities.

How To Manage An Effective Team

Much has been written on the management of teams; even so, many questions on managing teams in an effective manner still remain. How much freedom does a company give them? How much control? How much bureaucracy does one put in place to insure quality designs? How much day-to-day management to insure the product team stays on schedule or does not take too large a technology risk? These are all questions that management must answer. Some insights into these answers are discussed below.

1) The Team Should Function Under Its Own Leadership. The fundamental nature of a team requires it to operate under its own leadership. A team needs to have the freedom and autonomy to manage its own processes given some of the guidelines mentioned earlier. If a senior manager must operate

Concurrent Engineering Teams 85

the team on a day-to-day or a week-to-week basis, then the team is not operating as a team, but as a part of the manager's staff.

2) Provide Goals For The Team To Achieve. Effective management of a team requires the establishment of goals for the team to achieve. These goals should be realistic and achievable within the time frame of the product development. The goals should be targets on market penetration, revenue targets, product cost, functionality, expected volumes, and project schedule. The measurement towards these goals are a major item in each product review by management.

3) Encourage Team Collaboration. Collaboration is a key company environmental element. Functional managers, team leaders, engineers, and team members should encourage it to the highest extent possible. Collaboration means dependency on others for success. Thus, the goals and individual measurements must be both separate and combined. For example, a professional athlete may get a large increase in pay based on outstanding performance, even though the team had a losing season. However, this athlete did not get the bonuses based upon team success, such as a playoff bonus. This dual reward scenario is done because excellent teams have to be built over time. This reward structure provides a long term view of encouraging collaborative team efforts while retaining dependable individual performance. Just as in the case of the professional athlete, new product team members who have contributed greatly must be rewarded, even if all goals are not achieved.

4) Measure By Planning And Reviews. A team's contribution to the organization prior to the completion of the product can be measured by the quality of their plan and by the information provided at the review. These are key measurement points for both management and the new product team. Adequate plans are the first step in excellence; good reviews show that the plans are being achieved. However, the freedom of the team to understand the market and develop a product that fits this understanding should not be compromised by the need to follow the specific detail in the plan. Teams should continuously follow the market requirements and make sure their product will achieve the penetration identified. This is the ultimate success or failure of the team.

For instance, in a new product design review, the question of achieving the product development schedule or slipping the schedule is of major concern, especially given the impact of schedule slips on the business financials. The decision to change the schedule is an excellent one only if the planned product volumes will still be reached or exceeded. This can happen when market research shows the possibility of expanding the product into new markets and that slippage is needed to allow adequate time to develop the new features or options required. However, if the slippage is due to the lack of coordination, or the lack of collaboration, or poor execution, or just major misplanning, then it should be flushed out by the review process. These causes need correction by management since it is likely the product will not achieve the goals established due to the same underlying reasons which caused the schedule slippage.

5) Encourage Organizational Support. New product team members often find themselves viewed as outsiders within their own functions, since much of the team effort does not depend on the normal everyday functioning of the organization. Because of this, it is important for the senior functional mangers to outwardly support the team members and their participation in the concurrent engineering team. Managers must insure their organizations consider the new product activity at the appropriate level of priority, and factor the product information into their operation's activities in a constructive manner.

How To Preserve Expertise Of Your Key Members

Teams require the development of ideas through the contributions of its members. Original ideas are usually modified until they meet the needs of the majority of the team members. This provides for a better balanced solution, but if it's not done carefully it also can provide for non-excellence in the product offering. Teams that try to facilitate discussion and look to improve every contribution, and satisfy every need, invariably drive out excellence. Those that contribute get discouraged with their ideas being changed, and tend to contribute less. This is especially dangerous when experts are needed. Experts have a highly developed sense of the problem and the solution. This is usually developed over many years of work in addition to a strong educational foundation. It is unlikely that other team members will have or can attain this in-depth level of understanding. Because of this, engineering experts can easily become frustrated with team interaction and team critique, especially when they need to educate a team member about the basics of engineering expertise before they can begin to discuss the problems and the solutions.

Since expert review is difficult within the team, a foundation for the support of individual experts must be established if the team is to succeed. What is needed is the development of commonalty to make it easier for individuals to relate.

Several methods exist for the development of interfunctional commonalty. One such method used by several companies is to have each group spend time within the other's organization, actually doing the job or working alongside of an individual. This task should be long enough to give the other person an appreciation of the difficulties and constraints of the job. This might mean a day spent building and testing products in manufacturing, or a few days trying to develop lead generation activities in marketing, or several days in engineering trying to figure out from scratch the looks of a part or the technical analysis of a complex assembly.

At the beginning of the development cycle, and at significant intervals during the development period, time should be devoted to commonalty establishment between team members. This establishment provides a basis of respect for the specific expertise each individual brings to the team.

This does not solve the problem of technical experts who are brought in to provide their contribution to the team. This is especially true for a complex technical problem being addressed by a highly skilled engineer. One option to preserve this level of expertise in the product, is to treat these issues at the sub-team level. At this level, individuals with the appropriate expertise can discuss the technical issues at length, remain focused on the solutions and provide well thought out recommendations to the product team; thus, avoiding the situation where the team is forced to naively critique the expertise of highly competent individuals. If the team has a conflict with the recommendation and the team goals, the sub-team should be the one to address this with clear problem guidelines defined by the product team.

How To Overcome Some Common Team Difficulties

Teams are very complex groupings to manage. The skills needed to manage a new product team are not developed within normal functional organizations in industry. Typically, these skills are learned by trial and error. Thus it's important to understand the basics. In many cases, the teams will be the only place that the contributions of the individual members will be recognized and even understood. Thus, the motivation for contribution must exist within the team process itself. If the team process degenerates, it can greatly discourage the product development activity and undermine the collaborative environment needed for a successful product introduction.

Two fundamental team processes exist; the first is the team meeting and the second is the communication activity that occurs between meetings. Planning and execution are tasks done within these team processes. Both processes done effectively are vital for the healthy interaction of team members.

Team Meetings And Group Interaction

To keep the team meetings and group interaction processes effective, several fundamental steps should be followed. Good disciplined meetings allow all the participants to effectively use their time. Meetings that get bogged down in detail or are ill planned or members who don't come prepared waste the team's time and eventually lead to product problems. Meetings where the decisions made, milestones accomplished, and action items assigned are not written down cause the members to continually debate issues, remake decisions, and not accomplish important team tasks. In order to avoid these problems, specific techniques have been developed. These are outlined below. Many of these steps should be done by the team leader, but they can also be used by the team members when appropriate.

> 1) **Agendas.** Agendas should clearly identify the subject, identify which team members should lead the discussion and the amount of time allocated. Agendas should be published 3 to 5 days prior to the team meeting.
> 2) **Minutes.** Any team meeting should have minutes which record the decisions, the reasons behind the decisions, and the action items. The minutes should be sent out within 2 to 3 days of the meeting. The action items should list the action and the owner of the action.

3) **Between Meetings.** Key issues that need to be prepared for the next meeting should be discussed in detail outside of the meeting format. Small meetings (3-4 people) should be used for collaborative development between meetings.
4) **Issues.** Key issues that need to be worked or resolved should be converted into clear action items with clear owners. Confrontational issues should seldom if ever be brought up for resolution if they have not been worked between the two or three individuals having the conflict ahead of time.
5) **Developing Trust.** Team members must believe that the team processes will treat them fairly. Thus, there must be consistency in setting expectations, and in achieving expectations. Team members must develop a trust in each other's ability to deliver on their commitments. Criticism must be used sparingly, especially if it hasn't been raised to the person ahead of time outside of the meeting. Errors should be acknowledged and corrected rapidly.
6) **Check The Process.** At the end of each meeting check the process, i.e., has each team member provided feedback on how he thinks the process is going. Record the comments. Constantly strive to improve the meeting and interaction processes.
7) **Use One-On-One Meetings.** Many subjects will require one-on-one meetings with team members. Some members will be "stuck" on certain ideas that are not appropriate to implement, or conflict with the goals of the team. These should not be ignored, since these "hidden agendas" will create many problems if left unaddressed. Skillful team leaders will look for these underpinnings and work to establish meaningful solutions acceptable to all team members.

Concurrent Engineering Teams

8) **Focus On Information Dependencies.** Successful teams recognize the importance of shared information. They articulate their needs from one another. They understand the dependencies of their own individual work on the information provided by each team member. That's why the team organization structure works. Successful teams focus their efforts around the collaborative information needs from each other.

9) **Brainstorming.** Some solutions will require significant creativity and input from all the team members to resolve. Brainstorming is one team technique useful for this. In brainstorming, all ideas are recorded no matter how wild they may seem. Criticism is not allowed during this phase since it discourages idea generation. At the end, the common ideas are categorized and the themes established. The top three or four solutions are identified and discussed in detail. The team then decides which solution or solutions to pursue.

Collocation Versus Distributed Team Members

One question that invariably comes up is should team members be collocated (sit physically next to each other) or be distributed (with their home organizations)? Teams have worked both ways, so the real question is what is the most effective method? To answer this one must understand the current organization. Basically, collocation is intended to develop a trust relationship between team members. However, if the team members are away from their home organization, their access to diversified technical inputs is diminished. From a technical standpoint it is better to keep the team members where they have access to their day-to-day technical and data resources, unless relationships will not develop to the level of trust and understanding needed for new product development. For example, if the support organization has an unsupportive environment for new products then the answer is to collocate. Oftentimes, collocation is needed in the beginning. After the organizations adjust to the CE environment, then it is no longer needed.

Understanding The Team's Limitations And Implications For Success

Teams bring newness and change and have been known to fundamentally change businesses at their core. Effective teams are very dependent on the skill of the players selected, their support and management structure, and the willingness of the individual members to work as a cooperative entity. These structures have limitations that are important to recognize.

Skill Versus Role Confusion

One team limitation is the fact that skills and roles are often confused. Since the team is formed from different organizations and expertise, team members do not readily recognize or appreciate what makes an individual skillful in their function. Inputs received via the team meetings are taken as highly accurate and essential for that function. There is no way to test the team member inputs without causing a schedule delay and discrediting the individual (a team destructive technique). Thus, even though the input may actually be inaccurate, subjective, and non-representative of the function's true need, or simply just wrong, it will be treated and acted upon as correct. Furthermore, even when inputs are correct, a skillful presenter arguing for a minor change may outweigh the less-skilled influencer who has a much larger and more impactful change. Thus, in areas of disagreement, it is unlikely that effective and beneficial interfunctional compromises can be made. The end result of serious interfunctional disagreements is normally decision making by the team and management leadership. Thus, teams are very dependent on the skills of the individuals and their leaders.

This lack of recognition of experts in their function is an important limitation since it has the potential to result in conflict that is damaging to the success of the product and the business result. For example, a marketer knows the true value of product literature with simple messages, whereas the design engineers want to tell the customer every piece of technical information. The same functions have difficulty agreeing on product features versus the customer's willingness to pay for improvements.

Expertise Logistics

In the area of manufacturing, the problem of the new product team member is one of finding the right expertise to bring to the problem. Expertise in manufacturing is usually in a narrow band; for example, sheet metal bending, or printed circuit board testing. Thus, many individuals need to be brought to the development team at various points in the cycle. Once this information or expertise is brought to the team it must be sustained, often without the presence of the manufacturing expert. Since slight design changes can have major manufacturing impacts. This role normally involves multiple people and multiple inputs and is logistically and technically difficult. In order to use this narrow band manufacturing expertise in a concurrent engineering environment, organizational structures need to be established to support the input into design. Electronic communication greatly facilitates this activity.

Management Limitations

Management also imposes team limitations that affect the operation of the team. Two examples are:
1) Technical background of the people selected - People use what they know. For example, experts in sheet metal will rarely recommend plastic alternatives. Teams with the skills to reduce assembly time will focus on this, and teams with the skills to minimize fabrication costs will do that. Skill and background of the team members provides a very important focus for the team, and the results are almost predictable.

2) Size of the team and the support teams - Small teams seem to always work since the closeness can overcome some of the obstacles, whereas large teams are overwhelmed by the inability to interrelate, given the diversity of expertise. Different structures are required depending on the size of the team.

Team Self-Imposed Limitations

The lack of recognition of expertise, confusion between roles and expertise, inappropriate background to make the needed contributions, large teams, constant exposure to criticism due to design process immaturity all leads to the frustration of the key engineers upon which the actual design rests. An important conclusion of understanding the team limitations is as follows. Given the patience to develop a product using concurrent engineering the first time, in which gains have leveraged the expertise of the team, subsequent products by the same players often do not have the same far reaching impact. The gains based upon their knowledge have been made and new knowledge sources need to be found. The same team may work well together and be very efficient, but pursue sub-optimal solutions. Previous compromises have set limits on the innovation and progressive risk taking needed in achieving great designs.

General Team Limitations

When talking about teams it is also important to understand when not to use teams. Teams work best when there is a need for a collection of expertise that is not common within a single organizational structure. However, organizational structures such as organizational functions and sub-functions work best to focus specific expertise and avoid duplication. Here are three examples:

First, team processes are slower than individual processes; thus, if each team member's contributions are not vital to the activities of the other members, then the team process provides no real benefits.

Second, teams are limited as a management structure since by their very nature it is difficult to recognize the contributions of individual members on the result. For example, long term research and development over successive years would be very inappropriate as a team structure due to this lack of recognition.

Third, team structures are also limited in the management of operational processes. Operational processes require efficiency and dedicated expertise for success. Organizational structures that use delegation, structured organization, repetitive processes and can measure short term progress through metrics are better suited to functional organizations.

Example Of Industry Teams

An example of a concurrent engineering team is shown below. The core members each have their own sub-teams. Often, sub-team members are shared across multiple projects. This team is also depicted in Figure 15.

Engineering Team Leader

Lead Engineer
    Electronic Consultant
    Electronic Engineers
    Mechanical Engineers
    Software Engineer
    Industrial Designer
    Human Factors Engineer
    Power Supply Engineer
    Regulatory Specialist
    Diagnostic Engineer
    Component Specialists
    Designers/Detailers

Manufacturing Engineer
    Process Development Engineer
    Technology Specialist
    Manufacturing Test Engineer
    Cost Analyst
    Materials Specialist
    Supplier Interface
    Quality Engineer
    Prototype Developer
    Production Planner
    Component/Vendor Qualification Engineer

Marketing Manager
    Marketing Research Manager
    Communication Specialist
    Sales Support
    Competitive Analyst
    (Key customer contacts)
    Distribution Channels Specialist
    Industry Marketing Specialist

Engineering Services
    Qualification Team
    Testing Services
    Laboratory Services
    Reliability Engineering

Information Technology Services
    CAD Specialists
    Network Specialist
    Database Application Specialist
    Applications Experts

Product Support
    Field Repair Engineer
    Field Repair Planner
    Documentation Specialist
    Training

CE Mentor

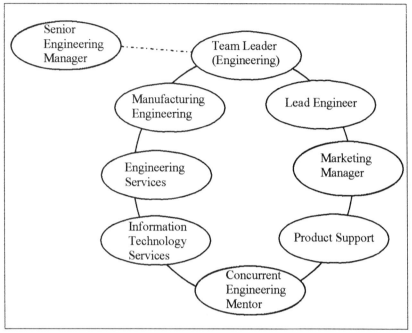

Figure 15 *The core team members are shown above. They use an extended staff and support organization to provide the activities needed. This team meets on a regular basis to discuss the status of the activities, and problems, and review their progress against the project schedule.*

Summary

This chapter provided a detailed look at the roles and responsibilities of new product team members. It developed comparisons of concurrent engineering teams to sports teams, and standard teams used within industry. It identified an effective method for establishing a team within a company and identified the roles of the senior manager and the team leader, as well as the roles of the key team members, including design, marketing and manufacturing.

Furthermore, an outline for an initial team plan was provided. A description of the content required by organizational functions was defined. This is the critical content required to support the concurrent engineering team environment within the company. How senior managers should set goals, and measure and manage a concurrent engineering team was described. Specific problems of team management were identified and addressed including identification of some of the limitations of teams. The key operating processes of the teams were described, along with a few do's and don'ts. Lastly, a listing of the membership of a typical engineering team was provided for comparison.

Chapter 4

Selection Of Key Techniques And Methodologies

A significant piece of the concurrent engineering process is the proper and efficient use of key methodologies and techniques. Many of these have been developed in the last fifteen years. These evolved in order to fill missing gaps in the design process itself. They provide a way to focus some of the inputs given to design by team members and others. Some techniques and methodologies provide very specific numerical answers, others provide a way of comparing subjective data. Techniques and methodologies have emerged as a more reliable practice in facilitating key decisions in the design process. Significant research has validated the more progressive ones. This chapter discusses some key techniques and methodologies useful in the development process. Those chosen are the ones that support the idea of collaborative development.

What Are Techniques And Methodologies?

Techniques are simply a specified manner used to accomplish a task. Techniques by themselves are not necessarily

Key Techniques and Methodologies

progressive or good unless they are proven by data to provide consistently improved results. In music, a good technique is easy to spot from a bad technique. It is intuitive and obvious. However, in engineering, because the results are not as quick, good techniques are more difficult to identify. The intuitive processes that evolve into good techniques take years to formulate. Methodologies on the other hand are significantly more structured. Methodologies are defined as an orderly arrangement applying the principles of reasoning to scientific and philosophical inquiry. Methodologies in engineering are structured reasoning, often with numerical comparisons or calculations that have been proven successful through use.

What Specific Methodologies And Techniques Relate To The Concurrent Engineering Process?

Several methodologies and techniques are useful in the development process. Some provide insight at the concept level, others are very detailed and help at the end of the design process. The concurrent engineering process described in this book is a technique by itself. In the beginning of this chapter is an overview of several useful techniques and methodologies with descriptions. It is not intended to be an all encompassing list, but highlights a key sampling of techniques and methodologies useful in the concurrent engineering process. Detailed examples and further descriptions of these techniques and methodologies are given in the second half of this chapter.

1) Quality Function Deployment (QFD)[1]. This methodology provides a method of rating the importance of specific product features using customer input. These features are

---
[1] Materials in this section and in the subsequent section on QFD examples are from *Quality Function Deployment: Integrating Customer Requirements Into Product Design*, edited by Yoji Akao. English translation copyright (c) 1990 by Productivity Press, inc., PO Box 13390, Portland, R 97213-0390, (800) 394-6868. Reprinted by permission.)

then compared to competitive products. Through these comparisons, the intended new product can be analyzed against both the customer requirements, and the competitive products currently being sold to fulfill this market need. A justification for establishing key features and a targeted product pricing is a conclusion of the QFD process. As a result of the QFD, the team has a clear feature priority, and an indicator of the price this feature combination may yield in the marketplace. QFD is used at the concept level of the design process, and helps to focus design attention on the key customer demands and competitive differentiation. It eliminates those features of low priority considering them wasteful.

2) Design For Manufacturing (DFM). This is a method of design with the goal of understanding the product's future manufacturing processes during the design stages. As a direct result of this understanding, the team can minimize manufacturing costs, and maximize manufacturing quality. Several methods and techniques have emerged as a way of implementing DFM. In its simplest form it represents design rules and guidelines that the design engineer can apply, often sitting side-by-side with the manufacturing engineer. In more complex approaches a structured procedure for rating an assembly and its components is used. This rating identifies a comparison of product complexity often using assembly time or fabrication time as the common indicator. From this time rating the design engineer determines fabrication, assembly and fastening methods as well as the need for combining parts. In even more complex implementations a combination of simulation, manufacturing expertise, rules, and assembly analysis is used to provide a low cost manufacturing strategy to follow throughout design. DFM can be used throughout the process by selecting the approach, technique or methodology that fits the specific level of design.

Key Techniques and Methodologies

3) Cost Driven Design (CDD). At the concept stage of design a methodology called Cost Driven Design is useful in determining the specific changes required to be made by design. CDD is also useful in determining the process changes required to be competitive. CDD helps the team understand the cost elements prior to designing the product, and provides specific cost targets on a sub-assembly and piece part level. The methodology allows the team to understand existing products through teardown and benchmark techniques. From this method the cost challenges required to be competitive can be established and specific development programs identified. This method is especially useful in cost sensitive products.

4) Design For X (DFX). This is a concept that includes many methodologies. It emerged as a team goal where various team members such as the quality engineer could offer their contribution to the product. It helps team members focus their unique needs where "X" fits these needs: for example, design for reliability, design for serviceability, design for competitiveness, design for testability, design for manufacturability, design for quality, design for safety, design for internationalization, design for maintainability, design for compliance, design for assembly, design for analysis and others. "X" usually represents a set of activities performed by the team member responsible for the specific function. This team member uses structured procedures based on previous products, and often utilizes software tools in the analysis process. DFX is useful throughout the process to insure key areas are highlighted and addressed appropriately. It forces individual team members to focus on their unique contribution to the design process, and utilizes the functional organization to develop design guidelines, methods and software tools to support the design analysis activity.

5) **Pugh Process**[2]. This is an analysis method of comparing various alternatives. It is useful in identifying which alternative attributes are the better attributes. It helps the design team gain greater insight into the strengths and weakness of potential solutions. Often during the use of this analysis an improved alternative containing the majority of the better attributes emerges. Pugh processes are useful as a method to select alternative concepts and designs. It is especially useful at reviews to identify why specific selections were made over other alternatives.

6) **Taguchi's Robust Design Approach**[3]. This is a method of design which establishes a specific approach to design. The objective is to minimize the deviation from the desired target level (for example, a power supply output voltage), while minimizing manufacturing costs. These deviations addressed by the robust design approach are typically due to three areas: manufacturing process (for example, assembly variations/errors), environment (for example, heat transfer), and material variations (for example capacitance variation). The objective is to minimize the impact of these deviations on the designed function.

Robust design provides a three step approach to achieve this objective, with the first step being design development. The second step addresses the design parameters, and the third step addresses the tolerances. In the parametric design stage (second stage) the best combination of nominal values is sought. The assumption at this phase is the lowest cost components with the widest tolerances are selected for the design. Typically, several combinations of components are examined to give the best result which minimizes variation.

([2] Materials in this section and in the subsequent section on Pugh examples are reprinted by special permission from *Total Design Integrated Methods for Successful Product Engineering*, Stuart Pugh, Addison-Wesley Publishing Company, Reading, Massachusetts, 1991)

([3] Materials in this section and in the subsequent section on Taguchi methods and examples are reprinted by special permission from *Quality Control, Robust Design, and the Taguchi Method*, Khosrow Dehnad, Chapman and Hall Publishing, New York, New York, 1989)

The third step is tolerance design. It is carried out only if the performance criteria is not met during the parametric design stage. Only the areas that significantly impact the desired target level by their variability are considered for tolerance tightening. Typically, components with tighter tolerances are selected one-by-one until the desired performance is reached. This technique is very useful in electronic design and some mechanical design applications.

7) Experimental Design Techniques[4]. Typical design analysis approaches examine the effect of varying a single factor while holding the other factors constant. Experimental design techniques, developed by Taguchi and others, provide a better way to maximize the information gained from a minimum number of experiments, while several of the parameters are changed from one experiment to the next. This method is especially useful when product characterization or process characterization is needed. In combination with robust design techniques it can be used to understand which process, environmental or material variations have the largest impact on the product functionality.

8) Design Stress Analysis. This technique uses methods of accelerating early life failures, and combines these with parametric stresses, such as timing variations. The purpose is to help understand the weakness in the design and the weakness of the components selected, thus eliminating potential failure modes before they get a chance to exist in the product. This is a design quality and reliability improvement activity, and is best used where designs are complex and reliability between design variations needs to be clearly understood. This technique combined with design of experiments provides the quality and

---

[4] Materials in this section and in the subsequent section on Experimental Design Techniques samples are reprinted by special permission from *Quality Control, Robust Design, and the Taguchi Method.*, Khosrow Dehnad, Chapman and Hall Publishing, New York, New York, 1989)

reliability information needed to decide between design alternatives. Additionally, this technique combined with simulation can be used to predict design function and shorten the final design testing stages.

9) Benchmarking And Competitive Analysis. The techniques used here establish the required design goals. The method is designed to functionally benchmark the current product against the competition to understand the current level of functional competitiveness. This data can often be found in published reports by independent consulting agencies or developed independently by the company. Additionally, product teardown is a technique for understanding specific component design goals around cost, manufacturing process steps, distribution strategy and other areas.

10) Rapid Prototyping. This methodology has emerged as a bonafide technique for understanding the specifics of the design. Early prototypes, such as those produced by stereolithography equipment can be manufactured to validate appearance, demonstrate concepts, and provide insights that only physical representations of the design can show. This methodology is especially good for early customer feedback and feedback from manufacturing. Many design improvements have resulted from the use of rapid prototyping. These range from major changes in the product design to minimizing process costs and preventing design mistakes. Additionally, early prototypes built out of testable materials can be used for design experiments to test areas of concern or to characterize the design. Prototypes that are produced in their normal materials but in a rapid cycle are of the most use since they can be subjected to a rigorous set of tests and feedback.

11) Customer Focused Design. This is a methodology that provides for a significant amount of customer input into the design process. It requires the building of samples and models that can be put in front of various customers for trial and usefulness assessment. The customers responds to these alternatives. The design team reviews the responses and retries the process with a different set of customers, and a modified set of product samples or models. This is repeated until "customer delighters" are reached and customer satisfaction for the offering price is clearly understood.

Selection Of The Appropriate Techniques And Methodology

Several important techniques were identified. Selection of the appropriate ones should be carefully thought out. All of the above techniques require training. Some require hardware and software. For many companies they represent change in current practices, and have the problem of overcoming the built in resistance to change in any organization. The selection of which ones to implement first depends on where the organization is in implementing concurrent engineering and in implementing these techniques. Selection of specific techniques should be based upon product impact and ease of implementation.

It is the organization's role to provide the training and the support for specific techniques and methodologies. These organizations should also be investing in developing these techniques and methods to specifically fit the company's product set. Even though it's uncertain which ones to choose first, they all are needed and a clear program to bring them into the organization should exist.

Examples Of Key Methodologies And Techniques

The remainder of this chapter describes in more detail the specific methodologies and techniques needed for concurrent engineering. These descriptions are intended to give a brief understanding, not a comprehensive understanding. In almost each case an entire book has been written around the specific methodologies and practices described in this chapter.

1) QFD Quality Function Deployment

This methodology was devised by Japan's Professor Yoj Akao, and was brought into the United States in the late 1980's This process involves distinct steps with the results entered into a matrix for ease of comparison. These steps are:

1) Customers are surveyed to determine what they value in a product, and how they rate the leading brands against these values. The values are labeled Customer Demands and the ratings are labeled Customer Satisfaction. See Figure 16.

2) A Functional Characteristic Matrix is established that correlates the Customer Demands to the technical characteristics of the product. Each Functional Characteristic is measurable, such as voltage or time. See Figure 17.

3) The competitors products are benchmarked against the Function Characteristics and their product information is added to the matrix in the section labeled Benchmarks.

4) From the matrix established, it clear to see which Functional Characteristics will provide a competitive advantage in Customer Satisfaction. This is done by making sure the new design (benchmark section) equals or exceeds the competition in each Functional Characteristic which correlates to the highest Importance Rating.

These characteristics are studied for ways to improve them. These improvements provide the feature basis for the new product

5) The targeted improvements (New Design #2) are identified on the matrix in the Benchmark section. Through comparison of the Functional Characteristics to Customer Satisfaction, the targeted improvements can be rated as to their anticipated customer satisfaction.

6) From reverse engineering, estimates of the competitors costs are identified. Competitor's pricing is researched and used to identify competitor's product profit contribution. Market share is researched. This information is entered into the market section of the matrix. From this a determination can be made on pricing the new product. If the key customer features are exceeded then the product can be priced at a slight premium over the competition. If the company wants to grow its market share then the price can be set more aggressively.

QFD Example

All this sounds more complex than it is in practice. In order to explain it more clearly, let's take an example. Thermometer Inc. is an imaginary company that makes handheld electronic thermometers used in hospitals.

Step 1. Thermometer Inc. surveyed the users of its equipment and the purchasers of the equipment to determine what they value most in a handheld thermometer, and how they would

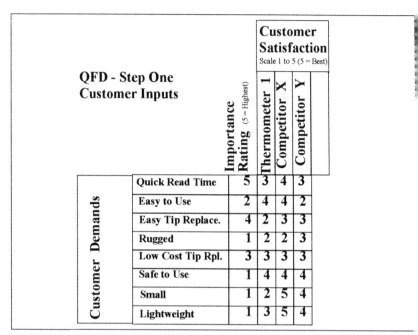

Figure 16 QFD--Customer Requirements. *Step one in the QFD process is to find out what your customers want and how they rate it in order of importance. Then customers are asked to compare your current product and the competitors product against that requirement. Each evaluation is given a rating 1 through 5 with 5 being the best.*

rate their current product, Thermometer 1, and Competitor X and Competitor Y. They received the following information (Customer Demands):

      Quick read time--start until temperature reached
      Easy to use
      Easy tip replacement
      Rugged
      Low cost tip replacement

Safe to use
Small
Lightweight

For each item, the customer included an Importance Rating and how they compared Thermometer 1, Competitor X and Competitor Y to these values. These items where entered into the QFD matrix. See Figure 16.

Step 2. The Functional Characteristics which have relevance to the Customer Demands were selected by the engineering team. Each Functional Characteristic must be stated in units of measure so it can be specifically compared. The specific selections are shown as follows:

1) Voltage at sensor--this is measured in hours above the operating threshold. This directly relates to consistency in fast response.
2) Human factors--device in left hand/probe in right, also look and feel. This is measured by user survey and rated as Excellent, Good or Poor.
3) Quick snap probe--time it takes to replace the probe, this is measured in seconds.
4) Probe cover material--this is measured as the rate of heat transfer, measured as the time to reach final temperature in seconds.
5) Total weight--measured in ounces.
6) Size--measured in height, and less than pocket size.
7) Case thickness--measured in tenths of inches, also drop height.
8) Storage design--measured in seconds to put away.

QFD Matrix

Functional Characteristics
△ Strong Correlation ◎ Some Correlation
▦ Possible Correlation

Customer Satisfaction
Scale 1 to 5 (5 = Best)

		Voltage at Sensor	Human Factors	Quick Snap Probe	Probe Cover Mat.	Total Weight	Size	Case Thickness	Storage Design	Importance Rating (5 = Highest)	Thermometer 1	Competitor X	Competitor Y	New Design # 2
Customer Demands	Quick Read Time	△			◎					5	3	4	3	5
	Easy to Use		◎	△			▦	▦	▦	2	4	4	2	4
	Easy Tip Replace			△					▦	4	2	3	3	5
	Rugged							△	◎	1	2	2	3	3
	Low Cost Tip Rpl.				△					3	3	3	3	4
	Safe to Use		◎							1	4	4	4	4
	Small						◎	△		1	2	5	4	3
	Lightweight					△				1	3	5	4	3
Benchmarks	Thermometer 1	4	Ex	38	45	14	6.2		.6	30				
	Competitor X	6	Ex	29	28	14	4.8		.6	28				
	Competitor Y	4	G	20	39	10	5.8		.7	26				
	New Design # 2	12	Ex	6	9	12	5.5		.7	12				

	Thermometer 1	Competitor X	Competitor Y	New Design # 2	
Market Price	86	98	92	88	$
Market Share	12	18	34	20	%
Profit	4	12	6	9	$

Figure 17 Quality Function Deployment (QFD). *This provides a way of extracting key features by listening to the inputs from customers, and converting that to competitive product advantage. The product can be compared to the competition on both functionality basis and a pricing basis. Thus, at the outset of the design process the price and functionality goals have been established based on marketplace conditions.*

Key Techniques and Methodologies 113

This information was added to the matrix under the heading of Functional Characteristics and correlated to the Customers Demands (see Figure 17). A triangle means strong correlation, a circle means some correlation and a square means possible correlation. If there was no correlation the matrix was left blank.

Step 3. Thermometer 1, Competitor X and Competitor Y were benchmarked for each Functional Characteristic. This data was added to the matrix in the Benchmark area. The competitors products were reversed engineered to see what the design team could learn about their design and spot competitive strengths and weaknesses.

Step 4. It became clear from the matrix that the customers wanted an improvement in both response time and tip replacement. These are both areas that would improve hospital productivity.

One design team was given the task of improving response time. This team knew that the response time was largely a function of the sensor, a part that was the same in all three thermometers. The key to improving sensor response was to maintain the voltage near maximum for a longer period of time. As batteries wear out, voltage degrades. This could only be achieved by adding another battery. They believed they could do this with only a slight increase in weight. Given the low ranking of weight in the QFD by customers it looked like a good trade-off.

Another team was given the problem of tip replacement. A new tip cover would be needed that could readily transfer heat and be simple to replace. Since these tips required replacement after each patient, quick connect and disconnect was a must. The team chose a two part tip; one part had a small copper metal piece that would conduct the heat from the tip to the sensor, and the second was a plastic cover. In the manufacturing process, the metal could be inserted into the plastic mold prior to plastic injection so that upon leaving the plastic mold it would be one piece. The metal tip could be made small and thin so as to rapidly transfer heat improving response time. Additionally the plastic could be designed for a quick twist on and off. The material chosen could withstand normal sterilization temperatures for reuse.

A third team designed the shape and size including a large readout, and a spot to store the probe when not in use.

Step 5. These improvements were defined in the requirements for a new product called *New Design #2*. This product's key feature information was added to the matrix in the Benchmark area, where the product's Functional Characteristic could be technically specified. Then, using the correlation symbols, the new product's Functional Characteristics were related back to Customer Satisfaction. An anticipated Customer Satisfaction rating was added to the matrix for New Design #2. can be seen in the QFD Matrix that the new design would exceed the best designs in key customer areas. The competitive advantages were obvious.

Step 6. Competitor pricing was learned through sales literature. The competitor's products were costed out by the manufacturing team member during the reverse engineering process. Price and cost were compared to determine profit contribution and this information was added in the market section. Market share information was researched from trade publications and market research firms and added to the matrix.

Next, the pricing for the new design was determined Thermometer Inc. could choose to charge a higher price based upon the added functionality, but instead chose a competitive price in order to increase market share. This was especially important since hospitals do not reevaluate suppliers of these small tools very often. Having market share meant significant repeat business in thermometer tips.

2) Design For Manufacturing

Design for manufacturing (DFM) has been evolving in the United States over a period of years. It originally started out as a producibility function, where the goal was simply to make sure a given design could be produced in volume in manufacturing. This evolved as manufacturing engineers realized that to control product cost in a meaningful way it must begin in design. This is true for incoming part costs, processing costs, yield costs, overhead costs, test costs, etc. For some designs, as much as 80% of the costs are set at the time of design. This means that manufacturing can only influence the other 20% with manufacturing efficiency, inventory carrying costs, overhead control, and other programs.

DFM methodology is multifaceted, but essentially can be broken up into three parts. The first part applies to the concept design, the second to the detail design and the third level is the final adjustments just prior to production. Each level has a finer and finer level of detail, and significantly different activities.

DFM At The Concept Stage--Cost Driven Design

The concept stage of design is the point at which design decisions and manufacturing inputs can play their largest role. This is the point at which the product cost structure must be established as competitive. Cost Driven Design (CDD) is a structured methodology that helps the design team understand where their cost competitive weaknesses are, identifies what cost improvements should be made to have a successful product, and identifies the specific cost targets. The process shown is one that the author has developed and used successfully. This is a simple multi-step process used at the outset of the concept phase of the design process, and is shown below.

Step 1. Identify the competitor's product and perform a teardown analysis. Cost out the competition's products as if you were producing them yourselves, using competitive sources for the parts. Thus, if the competition is producing a part overseas that you produce locally, use the overseas costs (assuming they are lower). Also identify where you anticipate the competitors to make cost reduction progress by the time your new product will be shipping. Enter this information into the CDD Cost Targeting Analysis matrix shown in Figure 18. In addition, in the small boxes, enter the performance comparison to your current product as: S for the same, (+) for better, (-) for worse, and (m) for major differences. Then enter the performance comparison of your new product to the competition's new product due out in the same time frame. The ideal goal is to end up with either better performance and a similar cost structure or similar performance and a better cost structure for your new product. Too many (m's) should be an indication of an unfair comparison of function and these should be reexamined. Next, estimate the product volume for comparison purposes and enter it into the table in Figure 18.

Step 2. Develop an estimated parts list for the new product. The parts list obviously will not be completely accurate when compared to the final product but it is necessary to determine cost problem areas in the design. If any unique options exist and provide competitive advantage they should be identified. The parts list is cost estimated using previous products as a guide. Where the design is significantly different a cost estimate is made by manufacturing. Both the current product cost data and the new product cost data are entered into the CDD Cost Targeting Analysis matrix in Figure 18. Only the key sub-assembly items should be analyzed, as the remaining parts and sub-assemblies fall under miscellaneous parts. The top part of the matrix (existing products and future products) is now complete.

Key Techniques and Methodologies 117

Step 3. Analyze the cost data by determining the difference between the competition and your new product. List both the worst case comparison and the best case comparison.

Cost Driven Design Cost Targeting Analysis		Product Cost (In dollars)						
		Complt Asmbly	Sub-Asmbly #1	Sub-Asmbly #2	Sub-Asmbly #3	Misc. Parts	Unique Options	Annual Volume (units)
Existing Products	Current Product	88	38	42	8			120k
	Competitor X	80 [+]	30 [+]	40 [s]	10 [+]	[-]		450k
	Competitor Y	86 [+]	38 [+]	38 [s]	10 [s]	[-]		350k
Future Products	Competitor X (new)	78 [-]	28 [s]	40 [s]	10 [s]			450k
	Competitor Y (new)	78 [-]	30 [s]	38 [s]	10 [s]			350k
	New Product	82	36	38	8			450k
Cost Target Analysis	Best Case	-4	-8	+2	+2			
	Worst Case	-8	-8	-4	+2			
	Selected Targets	**	**	*				
	Final Design	68 [+]	24 [+]	36 [s]	8 [s]			

☐ Indicates comparison of performance
(s) = same, (+) = better, (-) = worse,
(m) = major differences

Figure 18 Cost Driven Design--Cost Targeting Analysis. *This provides a methodology of understanding competitive costs and the changes needed to achieve these costs.*

Select the cost areas that are the least competitive. A double asterisk indicates a major target, a single asterisk indicates a minor target, blank indicates little or no major efforts. If the product is already competitive, then non-major areas of improvement can be targeted. The final design row is left blank until the remaining analysis is complete.

Step 4. The next step is to break down each of the selected sub-assemblies into more detail. This is done by using the Modification Analysis Matrix (see Figure 19). The parts list for the new product is compared to the parts list for the competitive product. Cost adjustments are made for unknown part costs, recording the assumptions. The differences between the competition and the new product are reviewed and specific parts

Cost Driven Design Modification Analysis		Sub-assembly #1 Electronics				
		Detailed Cost Analysis and Modification Selection				
		New Product	Compt. X	Delta	Change Category	Charac- teristics
Parts List	Part # 1	$8.00	$10.00	-$2.00	P S	Material Source
	Part # 2	$14.10	$4.10	-$10.00	D	ASIC
	Part # 3	$6.10	$10.10	+$4.00		Material
	Part # 4	$1.20	$4.70	+$3.00		
	Part # 5	$1.00		-$1.00	E	ASIC
	Part # 6	$2.00		-$2.00	E	ASIC
Change Categories: D = Design, P = Process, S = Source E= Eliminate						

Figure 19 Cost Driven Design--Modification Analysis. *Detail cost information is examined to determine the types of changes needed in the areas of Design, Process and Sourcing. Specific characteristics are determined for areas to target.*

Key Techniques and Methodologies 119

are targeted for cost reduction. In cases where parts have no corresponding parts by the competition they become immediate candidates for elimination.

Step 5. Next, the types of change to be pursued are identified. They are identified as D for design, meaning significant design change is required to achieve competitive costs, P for process, meaning significant process changes are needed to

Cost Driven Design Competitive Advantage Analysis		Comparison Category Rating 1 - 9 (9 = Best Advantage)			
		Volume	Process Technlgy	Product Technlgy	Other
Existing Products	Current Product	120k			
	Competitor X	450k		ASIC	Far East Mfg.
	Competitor Y	350k			
Future Products	Competitor X (new)	450k		ASIC	Far East Mfg.
	Competitor Y (new)	350k		ASIC	
	New Product	450k	Reduce Material & Test	ASIC	Far East Mfg.

Figure 20 Cost Driven Design--Competitive Comparison. *Competitive advantages are examined for learning and understanding. Future products usually need both technical and cost advantages to survive in the marketplace. Unique competitive advantages area are also targeted. As shown above the Process Technology improvement was added by the design team to give the new product a unique competitive advantage.*

achieve cost reduction, and S for sourcing, meaning the part can be sourced to a different supplier with few or no design changes. Some parts may get two or all three focuses.

In order to help determine the selection between D, P, or S, the Competitive Comparisons Matrix (Figure 20) is used. Any known technology, process or volume advantages the competition has are listed. This helps to focus where the significant changes need to come from.

Next, analyze each part to determine the key characteristics of the part that are driving the costs to be non-competitive. The goal here is to develop the cost target list manageable by the team member in order to make real inroads in cost reduction. The characteristics are identified and entered into the matrix in Figure 20. The intent of this matrix is to provide an understanding of the competitive advantage the new design will have including those advantages not currently in a competitor's product.

Additionally, if the change descriptions (ASIC, Far East Mfg. etc.) are not descriptive enough, each box can be rated 1 through 9. Thus, the product with the best advantage due to their unique ASIC design can be rated higher than the other ASIC designs.

Step 6. Cost Solution Testing is shown in Figure 21. This is an analysis to determine if the method selected will work to achieve the cost target prior to beginning the design. For design (D) changes it is a list of the data gathering and analysis steps required to understand if the design changes are likely to yield the cost savings. For process (P) changes this involves the flow charting of the exiting process and the new process and identifying the number of process steps and the amount of process costs eliminated. For each process step the material costs, the process costs and the yield costs are identified. The process flow should identify where design changes have reduced some of the effort within a process step. For sourcing (S) changes the flow chart

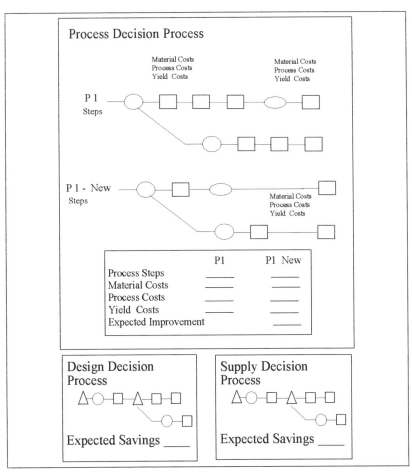

Figure 21 Cost Driven Design--Cost Solution Testing. *Each potential solution needs to be examined to determine if the ideas will work. The method for the Process, Supply and Design decision processes are shown above. The goal is to identify specific savings so the design team will have a strong understanding of where the cost reduction is likely to come from. These areas are identified and the target savings become the basis for the product design.*

identifies the research and vendor commitments needed to determine if the cost savings is likely.

Step 7. Once the Cost Solution Testing is complete the new cost estimates are factored into the cost for the new product. This data is entered into the Cost Targeting Analysis (Figure 18) as the final design. In addition, the performance data is entered into the smaller boxes, comparing the final design to the best competitor's new design. If the final design achieves the goal of better performance and similar cost or same performance and better costs then the analysis can continue. If not, then additional sub-assemblies and opportunities are examined until the cost targets are achieved. This final design information becomes the product cost budget for each sub-assembly.

Step 8. The last step is to pull together the tracking information to use as the design progresses and identify the collaboration requirements and product. This is a simple chart that shows current cost, projected costs, the characteristics which need to be changed, the approach (D,P,S), the resources and the collaboration required. Note the resources and collaboration may utilize outside firms. See Figure 22.

CDD Example

Using the case of Thermometer Inc., Cost Driven Design can be shown.

Step 1. Competitor X and Competitor Y products are put through a benchmarking and teardown analysis including cost estimating. The cost information is entered into the Cost Targeting Analysis (see Figure 19). It is noticed by the design team that the product labeling indicates that Competitor X's products are built overseas. This gives them a distinct cost advantage. Marketing has

Key Techniques and Methodologies 123

provided estimated volume figures from their competitive database and those are factored into the table. Competitor X also leads in the volume category. Manufacturing develops cost estimates taking this information into account and enters it into the CDD Cost Targeting Analysis Matrix.

Cost Driven Design Tracking Matrix

Sub Assembly	Current Costs	Future Costs	Characteristics	Approach	Resources	Collaboration
Electronics	38	24	ASIC, Material Testing	D / M / P	Outsource Testing	ASIC Design Precalibration
Enclosure	42	36	Snap Fit Part Reduction	D / P	Purchase Software	Vendor

Figure 22 Cost Driven Design--Tracking Matrix. *The analysis elements are pulled together in a reference matrix so they can be tracked as the design plans are made. Specific sub-assemblies have future cost targets, and specific sourcing, product, or process characteristics are identified for change. The resources required, and the collaboration required is identified.*

Next, the design team has learned through the trade press that the competitor's products under development are only going to be slightly changed, except for one subassembly. This information is factored into the competitor's new product cost data and entered into the matrix.

Step 2. The previous product, Thermometer 1, is used as the basis of comparison. Since the battery compartment is changing, there is an increase in cost complexity. The tip cover is expected to be simpler and easier to put on so this cost complexity factor decreases. The third area of change is human factors. The current design, although not very appealing was easy to produce, so the added human factors are expected to increase the cost of the product. A preliminary costed parts list is constructed based on the current product and the complexity factors are used to increase or decrease costs. The current product cost information on Thermometer 1 and the new product cost information is added to the matrix. The team is concerned since the new product cost estimate is still high (see Figure 18).

Step 3. The performance ratings are compared and added to the matrix. The team realizes that its older product has lower performance and is more costly to produce. The best case and worst case analysis is done on the matrix. The team identifies the cost of the electronics (sub-assembly #1), and the cost of the plastic enclosure (sub-assembly #2) as the two primary areas of cost focus.

Step 4. The team examines the detailed parts list for the sub-assemblies chosen and compares it to the most aggressive competitor using the Modification Analysis (Figure 19). The differences are recorded. The competition's advantages are categorized as design (D), process (P), or sourcing (S). The team identifies the key reasons for the competitive advantage such as ASIC design.

Next, the team determines the competitive advantage they want for their own product. First they record their understanding of the competitive advantages by rating the competition against themselves using the Competitive Advantages Analysis (Figure 20). The team reviews in detail the known technology advantages and process advantages that the competition had. As previously noted, Competitor X had a volume advantage. Competitor X also used a higher level of integration on its electronic components. They spend considerable time looking over and studying the competitor's product to determine the key characteristics that

Key Techniques and Methodologies 125

allow them to have lower costs. They conclude that the competition had failed to take advantage of any process advantages. The team sets D and S as targets to achieve parity with the competition. They further define process technology (P) improvements as their method of achieving a competitive edge.

Step 5. It is now time to test their anticipated cost solution. This is done in several steps. The manufacturing process is broken down using a technique of flow charting the manufacturing steps as in Figure 21. The team finds that for the electronics the cost is driven by the material purchased, and testing. For the plastic enclosure, the Thermometer 1 process is multi-step, including the adding of labels, four screws to hold it together, several complex assembly steps, and a thorough testing of the probe and calibration at the end of the process.

The team noted some key areas of process improvement. One was the long and costly calibration cycle. By purchasing the sensor precalibrated, a slight increase in part cost, the major step of calibration could be reduced to a quick on/off one temperature check. Additionally, tests could be automated if the volumes justified the expense.

Next, the team identified key design changes. The team noted that electronic integration both reduced the total component cost and the cost at test, thus making this an attractive option for the product team. With regards to the enclosure the team took the challenge of designing out parts such as labels and fasteners, and minimized assembly steps especially where features could be added to the plastics to allow simple snap in of parts like the Printed Wiring Board Assembly. Low cost connectors were added to eliminate the soldering of the probe wire to the board.

Then, the team identified the collaboration requirements. They identified that they could design the enclosure, if they purchased the mold flow analysis tools to help guide them in plastic design. However, they did not have the expertise or the software applications to develop an ASIC device. They decided to collaborate with a supplier in order to produce the ASIC device required to simplify the electronics.

Finally, given the above decisions the team set a cost budget for each main piece part. They agree that this budget must be achieved in order to have a competitive and profitable product. They filled in the cost budget information on CDD Cost Target Analysis Matrix (Figure 18) and compared performance. It looked like the new design would be better in both performance and cost when compared to the competition's new product.

Lastly, they pulled together the tracking information for future reference (see Figure 22), and identified the resource and collaboration needs.

DFM At The Detailed Design Level

Much has been published about DFM at the detailed level. In addition, many tools exist to support the manufacturing analysis, and provide important guides for the team. Some of these tools include:

1) Printed Wiring Board Manufacturing Guidelines where specific rules around spacing, board sizing, component types, etc. are provided to minimize costs.

2) Sheet metal costing tools which provide insight into the costliness of each process step, including stamping, bending, welding, grinding, plating, masking, painting, etc.

3) Design for test strategies which helps minimize costly testing and troubleshooting operations. Design for test strategies includes built in self-test, node separation for troubleshooting, and accessibility guidelines for electronic testers.

4) Mold flow analysis, which simulates the plastic flowing through the mold and identifies where problems may be created due to thin walls in the design, cooling of the material, etc.

Key Techniques and Methodologies

5) Machine simulations help to minimize setups, part changes on the machine, and drill patterns, identifying how the stock material will be removed, etc.

6) Design for assembly which is oriented towards defining the least costly assembly methods and minimizing the assembly effort required. This also helps engineers to identify which parts can be combined, and how to maximize quality by minimizing the assembly variables. Design for assembly analysis has many different approaches. Most look at the assembly sequence, analyze the fastening method for simplicity such as snap fit versus screws and calculate assembly time. Time is then the metric used as the minimization metric.

7) Process industries such as chip manufacturing have their own unique DFM rules and process capabilities that can be defined in documents or electronically. These process capabilities are formally reviewed as part of the design process. Violations are identified and decisions made as to the yield consequences. Well defined process capabilities will help to maintain high yields when new parts are introduced, thus keeping product costs low.

8) Database tools are useful in establishing communality between parts, and helping to minimize new part designs. Minimizing the number of parts helps to keep costs low in manufacturing, and helps to avoid inventory obsolescence. In addition, for the machining industry a fixtures database is particularly useful in minimizing fixture development and setup times. In the solid model approach the design and fixture can be electronically combined for manufacturing review.

DFM Rules And Guidelines Approach

For many areas of manufacturing there exist rules and guidelines as an approach to DFM. These rules provide both a training basis for the engineer and a checklist for his design. These rules are often generic in nature, and are often unstructured, requiring interpretation. Sometimes they conflict with one another, and in special circumstances they are incorrect a small percentage of the time.

Despite these difficulties, the rules and guidelines approach is a powerful one since it allows simple rules to be used which have broad impacts on manufacturing costs. Consider the rule "minimize the total number of parts". This rule affects much more than assembly or part cost. Each different part must be managed by purchasing or material managers in manufacturing. If the part is manufactured it must be planned through the process. In some cases fixtures need to be made. All this adds to manufacturing overhead. Thus, the total number of parts directly relates to manufacturing overhead costs.

Simple probability curves can be applied to this rule and show the impact of minimizing total part count on the ability of manufacturing to operate smoothly. More parts simply decrease the probability of all parts being available when needed. The result of this is inventory and extended manufacturing cycle time. The fewer the parts the better they can be managed (see Figure 23).

The same rule of minimization applies to process steps and the number of suppliers.

DFM At The End Of The Cycle

Some products have manufacturing features that are needed in order to process a part such as a holder that gets cut off just before shipping, or an orientation mark to facilitate assembly. These indicators or process tools get added at the end of the design cycle, and usually require a series of reviews with various parts of manufacturing and the suppliers to get the correct inputs for their specific processes.

Key Techniques and Methodologies 129

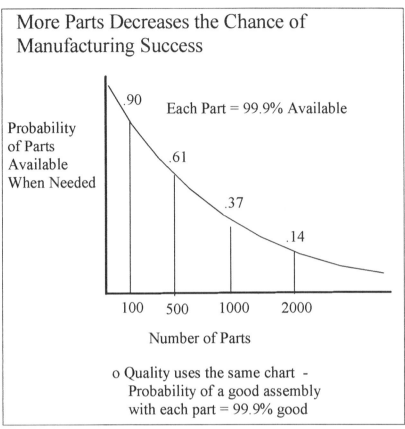

Figure 23 Rule 1: Minimize the number of parts, process steps and suppliers. *In manufacturing, the more parts that are created the more difficult it is for manufacturing to succeed. More parts are more difficult to manage even with high availability. In addition, the probability of finding a bad part at assembly increases. Simply put:*

> *More parts = More shortages/ material stoppages, more material management, more manufacturing variability, more quality problems at assembly, more suppliers, more design time, more assembly time, more management.*

Cost Accounting Impacts

One cannot consider costs as a design impact without running across the cost accounting dilemma of attributing overhead. One of the most common practices in manufacturing cost accounting is to apply overhead to hours of assembly. Thus, if the actual assembly time is one hour the actual cost is one times an hourly pay rate, for example, $10 per hour, then the actual assembly cost is $10. However, this number is multiplied by the overhead rate to give the final cost attributed to the product. Since the overhead rate is often three to five times greater than the hourly rate, this results in a cost of $30 to $50. It is discouraging to those who are developing cost reduction based on assembly time or manufacturing process time, since as they lower the actual process and assembly hours the overhead rate will have fewer hours to be prorated against and will go up. Thus, in our example, if the hourly rate is cut in half, the actual assembly cost goes from $10 to $5. The overhead rate nearly doubles leaving the final cost at $25 versus $30 originally. Thus, despite a good assembly time reduction, the total cost of the product remains approximately the same.

A concept called Activity Based Accounting is emerging which partially addresses this problem by applying the overhead costs to the activities which cause them. Thus, theoretically if an activity can be designed out then the overhead cost can be eliminated. Some of the activities which drive overhead costs such as the adding of new parts are given an appropriate cost penalty.

The problem of inaccurate accounting cannot be managed or impacted by the new product team as part of the development process. The team needs to drive towards the least costly method to manufacture regardless, and influence accounting practices to keep their products competitive.

Design For X

Design for X is a generic process. It allows the design team to pick certain areas that need attention and apply some rigorous process methodology to it to insure they will improve this particular area over previous designs. For example, reliability or maintainability may be identified by customers as an area that needs to be improved. The design teams decides that there are no quick fixes, such as a single change that can be made to improve this. They therefore establish design for reliability as a key item. In a strong concurrent engineering environment this item will be supported by the function, and they will invest time and resources in developing the rules and techniques needed by designers to improve their designs around this metric. Here is a commonly understood procedure for DFX. It includes the following steps:

Step 1. Identify X and identify a skillful team member who can focus on this area.

Step 2. Research commonly used methods of measurement. In the case of reliability, identify the methodology to calculate the product reliability, for instance, the mean time between failure (MTBF). Calculate the metrics for the previous product, and for the competitor's product if the data is available.

Step 3. Research rules used by competitors and others to improve the specific area. Identify design testing to validate the calculations, and identify the process changes required to improve this metric. For example, minimizing parts often will improve reliability, but only if the new parts selected are not significantly less reliable.

Step 4. Identify modeling tools to use to allow comparisons of various design alternatives. For example, solids modeling may be used as a method to understand ease of maintainability.

Step 5. Identify modeling and testing which needs to be developed by the function. Some of this may be long term development.

Step 6. Provide analysis, testing, modeling and feedback to the design team on improvements in the area of X.

Pugh Process

The Pugh process is useful when there are a number of alternatives to be considered, especially if the preferred solution may not be obvious to everyone on the design team, and is the source of debate. It was developed by Dr. Stuart Pugh from Scotland. The Pugh analysis like QFD and CDD is a matrix methodology. It allows the evaluation of each alternative based on a preestablished criterion. This matrix can be seen in Figure 24. The evaluation criteria are listed on the left. Across the top are the specific alternatives under discussion, with a brief description of each.

Also along the top is listed a Datum which is the basis upon which the other alternatives will be compared. For instance the Datum may be a competitive product, and all the alternatives will be evaluated against the competitive product. The team then tests each criterion against the alternatives. Each block gets one of three measurements. A "+" indicates that the alternative is better than the datum for that criteria. A "-" means that the alternative is worse than the datum for that specific criteria. A "S" indicates that the alternative is about the same.

The main point of the Pugh exercise is to develop a better understanding of the alternatives under discussion and how they will benefit the end result. As each alternative is discussed, notes

Key Techniques and Methodologies 133

and reasons for decisions are captured and agreed to. In many cases, a combination of the alternatives is developed so that it will receive a "+" for almost every criterion.

Pugh Process

Concepts / Evaluation Criteria	Concept Datum	Concept 1	Concept 2	Concept 3
	Description of Concept	Description of Concept	Description of Concept	Description of Concept
Criterion 1	Datum 1	+	+	s
Criterion 2	Datum 2	-	+	-
Criterion 3	Datum 3	-	s	s
Criterion 4	Datum 4	+	s	-
Criterion 5	Datum 5	s	s	-
Criterion 6	Datum 6	-	-	+
Number of +'s Number of S's Number of -'s		2 1 3	2 3 1	1 2 3

(+) means it is better than the datum based upon the criterion
(-) means it is worse than the datum based upon the criterion
(s) means it is the same as the datum based upon the criterion

Figure 24 Pugh Process For Concept Selection. *The Pugh process involves the selection of a Concept Datum. This is typically a competitive design that needs to be beaten. Several criteria are established to evaluate it. Then the remaining concepts are compared to the Concept Datum. This process is very helpful in sorting through which concept to chose.*

The +'s, -'s, and S's are tallied. If one alternative does not stand out, then the criteria are changed to achieve a less ambiguous result, or a more difficult Datum is selected. To make a decision, one or two top alternatives are picked and their weaknesses are examined. If their -'s can be changed into a + or S, then a new alternative can emerge. However, if they can't be changed, then the criteria must be tested to see whether they are severe enough to disqualify the alternative.

Taguchi's Robust Design Approach

The Robust Design approach is a powerful approach with its own terminology. Taguchi views design as a system, with inputs, outputs, control factors and noise factors (see Figure 25).

The goal of Robust Design is to use the control factors so that the noise factors do not change the response (or output). Control factors are defined as those factors which the designer can control and use to obtain the desired output. Noise factors are those factors which are not wanted and are not easily controllable. The noise factors can change the output. Some examples of the noise factors are the operating environment, the manufacturing process variations and material variations. Thus, in designing a power supply, the heat dissipation would be an environmental factor. The causes of low yield or fallout at test would represent the manufacturing process noise factors. Capacitance tolerance variation would represent materials variations (see Figure 25).

Another term used by Taguchi is the notion of Quality Loss. This is considered any deviation from the desired output. It can include losses to manufacturing, to the user of the product and to society in general. Under Taguchi's principles, the idea is to minimize quality loss and in this way minimize the total cost. Taguchi has developed a formula for this and two curves can be established (see Figure 26). One represents the quality loss of the user and the other represents the manufacturer's cost. As the tolerance is tightened the manufacturer's cost goes up, but the user's loss is reduced. Where these two curves intersect is the optimum tolerance. In Figure 26 this is labeled Target. The quality loss never goes below this point.

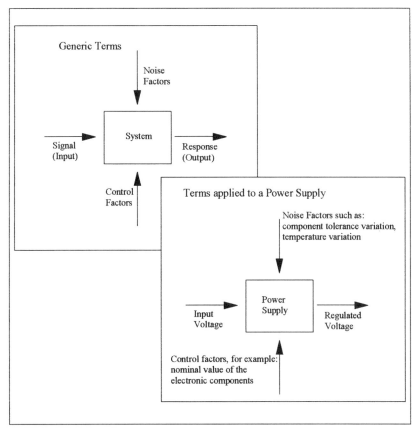

Figure 25 Taguchi's Terms. *Taguchi's terms can readily be applied to a power supply design problem. Minimizing the noise variation is done through selection of components which minimize the signal to noise ratio. For example, minimizing the capacitance deviation minimizes the influence of this noise factors on the output.*

The Quality Loss Formula can be used as follows. If Loss (L), in dollars, is known for a user failure (for example, $60.00 for the user to repair), and this failure is due to a specific cause y (for example a voltage drop to 90 volts), then k can be calculated $(60/(90-115)^2)$. A second equation can be established using the same value for k. Using the Loss Equation and the k value, y can be calculated for the product and a tolerance established (y-m). This assumes that a pre-established value for L can be set by the manufacturer's experience, for example, $20. Then, solving this equation identifies the value of y at the target point.

In addition to this, Taguchi advocates the use of Statistical Process Control (SPC) to insure the process variation does not exceed the tolerance established.

Taguchi's approach to Robust Design is a three step approach.

1) The first step is system design, where all the desirable functions are achieved. This step is the basic design effort of arriving at a workable system. Features are determined, functionality is set, material and components are selected, prototypes are made and tested, and so on. This is not different from the current design process in use.

2) The second step is parametric design. Investigation is conducted to determine the characteristics that affect loss and what their values should nominally be. The point in parametric design is to design with wide open tolerances in order to minimize manufacturing costs, while keeping the performance on target. The concept behind parametric design is to find the right combination that minimizes the design's sensitivity to the noise factors. Experimental Design Process is helpful here to provide the data to make these judgments.

3) The third step is tolerance design. This is the final step and requires the establishment of the most economical tolerances. Where noise factors are sensitive to tolerances, the tolerances are tightened until an acceptable level is reached per the Loss

Function. Tolerances are selectively reduced on the basis of cost effectiveness. Since tighter tolerances add manufacturing cost, only those tolerances necessary to achieve the desired output are tightened. In this way the total quality loss (any deviation from the desired output) is minimized.

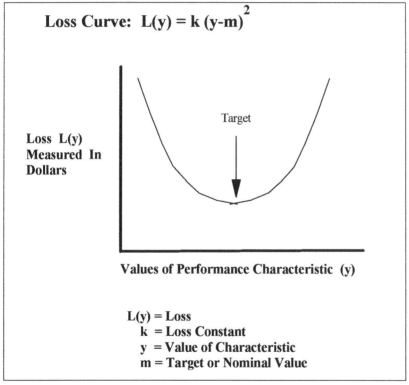

Figure 26 Taguchi's Quality Loss Curve. *Taguchi developed the Loss Equation to help relate tolerance establishment to both the cost of the manufacturer, and the loss to the users. Where the two curves intersect is the tolerance target. It minimizes both losses.*

Experimental Design Process

The experimental design process was used by Taguchi and others. It is a powerful tool to use to analyze variation using a pre-established set of experiments. The best way to explain the design of experiments is to use an example. If the design engineer wants to test the effect of process variation on the final design, a series of experiments can be established which will show the effect of various changes in the process on the final design. A matrix can be established which will allow the variation of all process steps and allow the engineer to determine which process changes have the maximum effect.

A matrix is shown in Figure 27. The row on the left is the number of experiments that need to be run (Trial Number). The columns across the top represent the manufacturing process (factors being tested). The numbers in the matrix indicate which process settings were used; two settings are possible per process. By performing this exact experiment, the process changes which correlate with the best product output can be identified. The precise selection can then be repeated for confirmation. If there is a discrepancy it indicates that some interactions are important but that more analysis and experimentation is needed.

Design Stress Analysis

Design stress analysis (DSA) is used as an early indicator of design problems. The idea is to overstress the design using accelerated life cycle testing methods, and combine this with parametric stress. Thus in the case of an electronic device, the test might contain rapid thermal cycling while adding timing variation to determine the stress effect the design is creating. The exact parameters for testing are dependent on the product and its operating environment, and are typically learned by comparing the results of current new products or know good products. This information provides both early life failures for the components selected as well as predicts design failures. Because of the severity of this type of testing its usefulness is often debated. Thus, the historical comparisons are important to develop since they help to understand DSA's ability to accurately predict future performance.

Orthogonal Array L_8 (2^7)

		Factors Being Tested							Results
		1	2	3	4	5	6	7	
Trial Number	1	1	1	1	1	1	1	1	
	2	1	1	1	2	2	2	2	
	3	1	2	2	1	1	2	2	
	4	1	2	2	2	2	1	1	
	5	2	1	2	1	2	1	2	
	6	2	1	2	2	1	2	1	
	7	2	2	1	1	2	2	1	
	8	2	2	1	2	1	1	2	

Matrix Numbers = Levels being tested

Figure 27 Design Of Experiments. *Designing experiments using this process maximizes the information obtained while minimizing the number of trials needed. The matrix experiment consists of a set of experiments where the settings of several products and/or process parameters are changed from one experiment to another. The sensitivity of the results to the changes made can easily be determined.*

Benchmarking

Benchmarking and competitive analysis are two separate activities. Benchmarking is a term with multiple meanings in industry today. First, from a performance standpoint it means to test a series of products to determine their performance against a prescribed unit of measure using a previously defined set of standard tests. For example, in computer benchmarking, the benchmark is how fast the computer will operate in terms of millions of instructions per second (MIPS). A computer with 300 MIPS is said to be better than one that can operate at only 200 MIPS.

Performance benchmarking is an important data point for all products in order to understand the cost/performance relationship of one's own products to other competitive products. Performance is based on some standard of operation. If one doesn't exist as an industry standard it is important to develop one for your product set. As a simple example, a coffee cup doesn't have an industry standard cost/performance metric. However, if one were to establish this metric it might be a measure of price/volume the cup held; thus showing the customer functional value.

The second area of benchmarking is the idea of benchmarking processes through the comparisons to known good processes in the industry or in other industries that have the potential of being adapted to your industry. This process is not an exact measure, but it uses observation and documentation. Because of this it is always open to subjective interpretation.

Competitive Analysis

The idea of competitive analysis also has many connotations. For the purposes of product development it means to analyze the competition's products. This can be done through industry reports. Trade shows are an important source of specific information about the competition's new products. If one wonders about the importance of trade shows to competitive information gathering, simply watch the interest at the new technology sections and the number of people taking and recording notes. Consultants also specialize in this type of information exchange.

If the competitive product is shipping it can be analyzed by reverse engineering. This is normally an excellent source of information for engineering, manufacturing and marketing. If your products tend to be behind the competition's then reverse engineering is the proper way to go. In reverse engineering a team of people (usually the team designing the competitive product) is assigned to slowly disassemble the product observing as many ideas as possible. These notes are then reviewed with a broader audience to understand the design, cost and quality implications. This data, although subjective, is often a vital source of idea stimulation and understanding the level of design excellence needed by your product to be considered competitive in the marketplace (see Figure 28).

Rapid Prototyping

This is a growing area of interest for many companies as the availability of rapid prototype equipment to new product teams is becoming more widespread. Several documented cases are in industry today using this technique. Chrysler corporation recently publicized the use of a stereolithography based rapid prototype process they used to develop a handheld tester. This tester is used by mechanics in analyzing fault modes when troubleshooting a Chrysler product at a service station. Chrysler developed an early model at the beginning of the process. This model developed by stereolithography included a false display and was weighted with lead to replicate its anticipated weight. This model was brought to six major dealerships around the United States. Dealer feedback and first impressions were recorded and brought back to the team to be discussed.

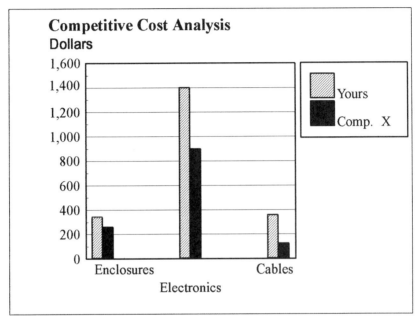

Figure 28 Cost Targeting Specific Sub-assemblies. *Comparing the cost data to a competitor's product is usually very enlightening. It helps the design team to understand if their new products really are competitive with other design teams from other companies.*

As a result of the early model, significant changes were made to the concept design. First, the shape was dramatically altered, the display was angled as a "stadium seating " type so it could be seen while sitting on the passengers seat during road evaluations. The display was altered to use a super twist LCD so it could be read in the sunlight (a super twist LCD reflects the sunlight improving its readability in sunlight). A hook was added so the device could hang from the underside of the hood of the

Key Techniques and Methodologies

vehicle, and a leather cover was added to make it easier to hold in the hand.

The next use of rapid prototyping occurred by using a plywood model for form and fit analysis. The results of this model showed the need for more internal space to fit the printed wiring board components (the unit had 5 printed wiring boards within it). In addition, more clearance for the LCD display board was added and standoffs were put in to improve mechanical support for the electronics.

A new SLA model was created and sent to the tooling vendor for evaluation. This helped to minimize the tooling cost.

The net result to Chrysler was a product done in 2 years versus 2.5 to 3 years done by previous methods. Further reductions in tooling, part savings and other development costs were seen. In addition, the product was much easier to use.

This demonstrates the typical use of rapid prototype technique. Many rapid prototyping methods are available, including wax models, foam core models, and machined plastics, to name a few. The importance of understanding the features early, as well as usability means that late changes in the design cycle can be avoided, and the resulting product will be more widely accepted by the customer base.

Customer Focused Design

This methodology is similar to the one described in Rapid Prototyping but more rigorous in its use. In this case, simple models are developed. The models need to be at a level where product use can be simulated. Customers are then asked to review these and attempt to use them in their normal environment. The environment is recorded, the usage process is recorded, and comments by the customer are recorded. This data is analyzed and used in the product alternative analysis. Renderings are also developed showing the product as it is used in the environment.

Lastly, after attempting to use the product, price ranges are tested with the customers to try and determine the real worth of the features and products. A clear distinction is made between features adding value (extra price) and baseline features (required for initial function). Any features that really delight the customers are highlighted. This process is repeated with several customers.

After the first research pass the data is reviewed by the team. Other models are made using these inputs, including all the baseline features, value adding features and customer delighters. Features of preference are not included but noted for possible configuration variations. These models are again tested with different customers in the same pattern, recording environment, usage, comments, and delighters. This process is repeated until broad based customer delighters are identified, and can be used as differentiators in the marketing and selling process.

Difficulties With CE Methodologies

New methodologies are not without their set of difficulties. For example, some techniques are widely acclaimed for their success by companies, but their true impact is often difficult to ascertain. Did the techniques used really focus the people to get the results claimed, or was the result due to other aspects of the development process, or due to excellent people applying a poor technique. The test of a good technique is the ease with which it can be understood and applied effectively, as well as the results achieved. The Concurrent Engineering techniques mentioned here have proven their value by many different companies.

A second problem with techniques is that they are not supported, funded, or available within many companies. Since many of these techniques cross the boundary between new product teams and supporting organizations it is unclear who budgets and supports them. If a development team chooses a technique such as QFD or DFM there are implications about information availability

Key Techniques and Methodologies 145

from the supporting organizations. Thus, to employ these techniques in a timely fashion by the development team assumes significant support from the appropriate functions. A common mistake by companies is to place the responsibility for the implementation of the techniques on the new product teams. This approach usually leads to non-success due to the time required or one-time success.

Technique usage by the development team has two limiting factors. One is the criticality of timing for using the technique. Obviously, a six month QFD effort has no place in a nine month development cycle. All QFD's aren't six months, but if the team does not understand the timing impact of a specific methodology, then the results could very counterproductive. The second is that techniques are emerging with little clarity about specific ownership and roles. Since change meets with natural resistance in industry, without clear ownership, techniques often have a difficult time finding their way successfully into the development process, despite the excellence they bring to the product and the process.

Summary

This chapter was a short review of some of the key methodologies and techniques that have added to the concurrent engineering process. This was followed by more specific descriptions and where appropriate, actual examples. Several sample matrixes were provided, as well as methodology diagrams. The methodologies included were:

        1) Quality Function Deployment (QFD)
        2) Design For Manufacturing (DFM)
        3) Cost Driven Design (CDD)

4) Design For X (DFX)
5) Pugh Process
6) Taguchi's Robust Design
7) Experimental Design Techniques
8) Design Stress Analysis
9) Benchmarking and Competitive Analysis
10) Rapid Prototyping
11) Customer Focused Design

Chapter 5
Selection Of Tools

Application programs, communication software, networks, frameworks, database products are all considered software tools of the new product team. The emergence of new tools has been a major factor in the emergence of concurrent engineering. This chapter describes some of the key tools each major team member should use as he participates in the new product development process. It describes the significance of the tool to the concurrent engineering process. The idea here is not to focus on any one company's tools or one tool versus another, but to give a generic overview of how the tool can be used to affect the overall process. After an overview, this chapter is broken down into the key team members' area of interest.

Software Tools Overview

One of the tremendous changes of the last decade is the improvement in software tools for the design engineer. These tools help capture design intent with the exactness needed to rapidly transfer designs into manufacturing processes. Computer-aided design (CAD) suppliers continue to improve their product's ability to aid the engineer in product development. For mechanical engineers the advent of solids modeling has allowed engineers to readily relate their designs to other team members, managers, suppliers, and customers, and rapidly respond to ideas and inputs received from them.

Additional software exists to help designers analyze what they have created and is commonly referred to as computer-aided engineering (CAE). For mechanical engineers and manufacturing engineers tools exist for structural simulations (finite element analysis), tolerance modeling and analysis, dynamic modeling, simulation of plastics and metals flowing through the mold, sheet metal bending/unbending models, part databases for access to previous geometric and dimensional information, numerical controlled (NC) machine process simulation such as surface milling, cavity milling, lathe, etc., manufacturing cost analysis tools, databases to analyze fixture usage and requirements and minimize setup, as well as producibility assessment tools and numerous others.

In addition to the specific tools, generic tools have provided significant help to the design process. For team members, and as general communication, electronic mail is especially useful. Messages are clear and written. Program direction can rapidly change and be understood by all team members and supporting organizations. When the teams are larger than a few people, a few organizations and/or collocation is not practical or possible, then electronic mail is a must in achieving timely and low cost design processes. The networking needed for electronic mail has an added feature of providing access to research, part information and historical data that might not otherwise be achievable in the time frame required for design development

With the introduction of information technology to the design process is the requirement to organize and manage its use. Product Data Management (PDM) tools provide the needed infrastructure for file control, i.e., design updates, version and revision control, authorization of design changes, file workflow, access rights to data, etc. In addition, specific expertise for various tools becomes a requirement to both help the users, and to create a better design and a better product.

Thus, tools have become a primary enabler for the concurrent engineer process. This enabler has brought another major shift to the design process that is not as widely discussed; specifically, a shift in the roles and responsibilities of the engineers, designers, specialists, analysts and team members. The electronic usage has shifted the idea of excellence and thoroughness in the design away from the engineer and into the "tool users". In some cases these are the same people. In others, there is a "tool expert" who can operate the tool but does not understand the design intent. Additionally, it has added "information transfer" as a design dependency, a step that is often not trivial. This "openness" in the design through tools and information access is a new source of design information release.

Another major enhancement to the design process made possible through both hardware and software advances is the concept of rapid prototype. The idea behind rapid prototype is to constantly evolve the design by reviewing several iterations of the design's physical representation. This continues until the design has met the criteria established by the full team, and the prototype is reviewed by customers.

Mechanical Engineering Tools

There are many tools that can be mentioned. Here are some of the basics:

2D CAD. This level of design is essentially drafting automation, and although it's been around for many years it is still being purchased by many firms. This level of CAD is entirely appropriate for non-complex designs and design documentation requirements. The two dimensions provide sufficient detail to describe the parts, especially if there is no need for downline processing of the information as in more complex assemblies. Some good examples here are cable drawings, technical manual sketches, or non-complex symmetrical parts. Additionally, 2D CAD is the starting point and a base skill for many 3D tools. For example, many designers develop a 2D part image and "extrude it" in software to form a 3D part.

3D CAD. Three dimensional CAD is in widespread use and is currently a very popular method of design. This method of CAD allows designers to see the outline of the parts in a "wireframe" three dimensional representation. Thus, corners, holes, special cuts, bends, etc. can be seen as lines on the screen. This allows the designer to capture the part as it was visualized by them. Another advantage of wireframe in the concurrent engineering process is that wireframe assemblies are see-through on the screen. This allows assemblies to be viewed for things like hole alignment, matching faces, etc. Thus, a more thoroughly validated model can emerge from the design process itself. One of the complexities of wireframe designs is that after a series of lines is added to the screen it is often difficult for others to look at and thoroughly understand. This is in part due to the optical illusion effect. In other words, is that line in front of or behind the others.

Surfacing. This tool allows surfaces to be added to the 3D wireframe model, thus the effect of seeing the part as a solid can be achieved, having similar benefits as mentioned earlier. Surfacing is not as robust as solids modeling (below) since both development of surfaces and changes made to them can be very cumbersome.

Solids Modeling. This is currently emerging as the most popular option for designers. This modeling method creates shapes on the screen. These shapes look like the actual part; even shading can be added to give a true picture. Using this method, designers can see what they are actually designing and can correct issues as they work. They better validate geometric shapes, fit of parts, clearances, etc. Also since it looks like the real thing an electronic design review can happen with other designers, manufacturing, marketing, management and even customers. Some companies send their solids model designs to their customers for review prior to spending significant amounts of money finalizing the design.

Solids modeling allows the assignment of more complex shapes to the design. Thus, the product can look nicer with significantly less effort. For these shapes to occur in wireframe a detail surface must be drawn in line by line. In a solids model, equations can be used to represent the solid shape. This is an important feature for the designer. Thus, the shape can be freely sketched and the dimensions can be specified as specific parameters or variables. These can then be easily altered to edit or create new designs.

Another important property of these modelers is the feature of associativity. This means that various geometric shapes can be associated with one another. For example, a boss or a hole placed a certain distance from an edge will remain at that distance when the piece is elongated, shortened, or changed in some other way.

In solids modeling, if any view is changed, the solids model and all developed views reflect the change.

For products where industrial design is important, solids modeling can be viewed as it is developed to determine how closely the industrial design renderings are being matched.

Many productivity enhancing features have been added by software suppliers, including those that automatically highlight interference fits. Solids models can be used by both marketing and manufacturing so that the shape and appearance of the design can be tested by customers and by manufacturing experts. Many solid modelers have an option of transparency, so the surface can be seen through, thus enabling designers to view the placement of components inside of the product.

There are many different types of solids modelers. Most are based on a concept called variable-driven modeling, where variables drive the shapes. Variables can include dimensions, equations, or other characteristics which define the geometry and when changed, alter the geometry. In addition, solids modelers can maintain constraints such as geometric relationships, algebraic relationships, or specific rules such as keeping two surfaces always perpendicular. Each solids modeler has different methods of developing the model and establishing these variables and constraints, and thus requires consideration when changes to the design need to be made. Some of these model types include explicit models, parametric modeling, variational modeling, and feature based modeling.

Explicit Modeling. A design defined only by its geometry is said to be an explicit model. More specifically, it is a model that does not capture the relationships between entities. In other words, the engineer draws the model and adds dimensions, and what you see is what you get. When changes are made, oftentimes substantial parts of the geometry must be redone. Thus, if a handheld device needs to have the contour redone, the entire geometry is likely to be recreated. In this modeling type some productivity is gained by developing standard shapes and recalling them for use.

Parametric Modeling. Some 3D solid modelers are parametrically based. The idea here is to make dimensions behave like mathematical variables (referred to as parameters). Thus, when a product is sketched the designer is actually generating a set of equations that the software solves to create the model. When changes are needed a new value is assigned to the parameter and the software recreates the model. In addition, these models require the designers to assign specific relationships to the elements as they are created, e.g., always at right angles, always 0.2 inches from the side, etc. These relationships are referred to as constraints. Each element must be fully constrained before proceeding to the next element. These constraints define how the model will behave when changes are made to a parameter.

Parametric modelers define geometric elements by capturing the constraint and variable history of construction. Thus, in parametric design the model captures the parametric relationships in a defined sequence. This sequence is followed when it recomputes the model. This computation method allows the rapid development of a model.

This ease of use characteristic has a downside. That is, since parametric models are sensitive to the order in which constraints are added to the design, when constraints need to be changed such as from a parallel constraint to an angular constraint the associated geometry usually ends up in an undesirable shape. As a result the geometry beyond the change point is usually deleted and recreated.

On complex designs, engineers typically scrap the old model and recreate a new one when changes are needed rather than attempt the difficulty of figuring out the constraint sequence which needs modification. This is potentially very destructive to a CE concept since valuable inputs can be lost during recreation; additionally it is prone to error. For this reason parametric design tools are often positioned as concept development tools.

Variational Modeling. This modeling method is similar to parametric. It allows the designer to specify geometric constraints, but also allows mathematical constraints to be included such as shapes, weight or structural properties. The chief difference is that variational modeling solves a set of simultaneous equations to define the solution. Because of this engineers only need to define critical relationships, and they do not need to constrain the entire model. Thus, when the model is recomputed the variational modeler takes longer since it requires more updating of the data. This delay is easily offset when a constraint needs to be modified. Unlike parametric models, variational models can be modified in any order and can remain underconstrained. This fits the natural way of working since engineers typically do not understand all the constraints at the time of design. In addition, variational modelers are not sensitive to the order of constraint placement. Thus, after the model is initially created, the variational model is easily built upon and modified without the need to recreate the geometry. This is a much preferred approach in a CE environment, it allows the construction of primitive designs, and the ability to add features, constraints, and more detailed geometry as the design process evolves.

Feature Based Design. Features in Mechanical CAD (MCAD) tools are emerging with a new importance for the designer. Form features are recognizable shapes such as holes, ribs, slots, bends, etc. They contain a specific set of geometric qualities connected by a specific topological relationship. Many of these forms can be parametrically created and modified. In addition, a feature can have specific properties, for example, a hole can be defined as a cylinder that is always a through hole, or always a blind hole regardless of future changes in part thickness.

An important advantage of form features is that they can easily be related to manufacturing. Feature libraries can be created that also specify the manufacturing process, by identifying the information required for each step. For example, a machined feature could be divided into the machining steps such as rough or finished, the machine parameters can be defined such as feeds, speeds, and depth of cut. Thus, the manufacturing knowledge can be encapsulated into the feature. By providing standard features between manufacturing and design, as features are selected, the manufacturing process and subsequent costs are known. This technique allows design engineers to make trade-offs based upon factual data. In addition, it simplifies the processing of data from design to manufacturing.

Selection Of Tools

Detailing. This tool allows the detailing of parts and includes dimensions, sectional views, design notes, and drawing symbols.

Parts Library. The parts library is usually defined as the repository for all completed parts. It also holds the specifications for all components currently used by the company. Thus, data bases can be searched to find both designs and parts to reuse. This parts library is where all functions can access the completed drawings. It provides the revision control, access rights, approval rights, security, etc. In addition to the company parts, catalogs of key vendors parts can be brought in through either networking or by establishing dedicated disks to the CD's supplied by the component vendors. These libraries provide the technical specification of the components.

Finite Element Analysis. This is a mechanical structural simulation tool. This tool breaks the CAD design into small sections called elements. Each surface is divided so that it looks like a wire mesh. The smaller the meshes the more thorough the simulation will be of that particular area. It is necessary to mesh the part because a surface is essentially an infinite number of points, and an infinite number is impossible to calculate. The computer recognizes each mesh as a single point or series of points, and computes the stress based upon this fixed number, thus the term finite element. Once the surfaces are meshed, then the engineer defines the forces and places them at the specific points and the load on the unit can be simulated for structural integrity. The viewing of the results is usually smoothed so that it appears as uniform shades of color rather than point specific. Multiple simulations are needed to represent various combinations, and to simulate anticipated problem areas. Finite analysis is also used to analyze vibration effects on the design. Finite analysis tools are powerful tools in evaluating designs. Other analysis types are under development which expand the capability of this type of simulation.

Dynamic Analysis. This tool type provides for a simulation of a part or an assembly to understand the impacts of applied motion and stresses on the structural integrity of the design. Combinations of dynamic analysis and finite element analysis can yield some good insights into the structural problems and vibration problems of an object or assembly in motion.

Rendering Tool. This tool allows the photorealistic renderings to be developed showing the environment in which the product is to be used, as well as sketches and choices regarding the design appearance itself. This tool includes easy methods to create 3D shapes, forms, and surface qualities and allows images to be combined with the design. These visualizations since they are electronic can be both shared by a wide variety of individuals, including those people doing the detailed design. With multiple windows a designer can have a rendering in one, and various views of the part being designed in others.

Information Library. This is usually defined as a library of documents such as regulatory specifications, UL requirements, IEC specifications, etc. that is accessible to the designer electronically. This allows specific specifications to be reviewed as the questions arise, rather than making a list of issues to be chased down at some future date.

Spreadsheet. This tool is very helpful for the many different calculations needed by engineers. This includes structural analysis, thermal budgets, etc. Cost tables should also be generated through the use of this tool.

Thermal Analysis. This tool simulates the heat dissipation of products, and identifies hot spots or thermal problem areas. These tools usually need to be validated with actual testing.

Design Optimization Tools. This tool allows the specification of variables, constraints, data points and goals. An example of this is minimizing weight, where the software evaluates many design iterations and display them as geometrical shapes, graphs or tables of data. This tool helps the designer think through the impact of changes in design variables.

Manufacturing Engineering Tools Supporting Mechanical Design

In general, there are many tools in use today for manufacturing engineers. For the most part they are oriented towards specific manufacturing processes.

Electronic Assembly Through CAD. Manufacturing engineers use CAD software to view the full system as it is electronically assembled together. By using this method of review, an engineer can look at multiple views, including the method of assembly, clearances, etc. This allows the engineer to provide input when the design is still easy to change. Typically, the manufacturing engineer is not allowed to change designs and is given read only files with an annotating capability (circle and comment in red). The master file is protected from change through password protect, product data management tools, or framework security. Companies such as Boeing have developed human CAD figures and placed them into the solids model to check access to critical areas for service and assembly.

Tolerance Analysis. This tool provides an analysis of the tolerances established, including statistical calculations on the effect of tolerance stack-ups.

Plastic Analysis And Mold Simulation. This tool simulates plastic injection into the mold, identifying problems of flow, cooling, and warping. This tool provides an understanding of the mold flow geometry, material properties, mold gate and vent locations, mold temperature and pressures. This tool helps to optimize the part design, and the mold design and minimize the cost of the mold. In addition, similar tools exist for cast metals.

Assembly Analysis Tools. This tool analyzes the assembly, starting with the structure of the design. It often defines the least costly way to assemble, provides a cost analysis using time as the metric, identifies parts that can be eliminated, and identifies tools and fixtures needed in assembly. Some tools simulate and develop the programming for robotic movement.

NC/CNC Programming And Simulation. Tools exist that automate file translation into NC programming language such that the result can be used directly by an NC machine. These tools simulate the machining of raw stock so that machine operations are known prior to use, avoiding any machine errors, including scrap calculations. In the process they simulate the milling tool paths, and provide insight for the manufacturing engineer. The simulated routes can be optimized to control scrap, bit ware, machine setups, material movement, etc. Milling software also includes sophisticated sculptured surfaces machining, and automated tool paths for turning, grooving, cavities and threading.

Sheet Metal Unbending Tools. This tool takes the CAD geometry for sheet metal parts and unbends the design to fit on predefined flat sheet metal stock. The software takes into account material deformation, bend radius, and material thickness.

Selection Of Tools 159

Rapid Prototyping. This tool allows the automatic conversion of data for rapid prototyping equipment. Rapid prototyping is used by the design team to understand the product in its environment. Early prototypes can be simple models. Some are carved from wax or clay to show the shape, other models are made by machining out plastics on a CNC machine, and still others are made from stereolithography equipment. This equipment builds a prototype in a liquid bath of photosensitive plastic. When a laser shines on a specific pattern the liquid hardens along the pattern. The depth to which it hardens varies by equipment and settings. Maximums tend to be in the 0.005 inch range. Thus, the part is dipped in the liquid bath and the process is repeated many times until a part is produced. Stereolithography requires the slicing of CAD objects. Each slice represents the laser pattern. The slicing is simulated so that it can be reviewed prior to building of the parts.

Process Simulation Tools. There exists a wide variety of specific process simulation tools. These tools help to identify capacity bottlenecks given the design, and process constraints that limit the design.

Manufacturing Process Files. Several companies use manufacturing technical files as a way of understanding the producibility of the design. The semiconductor industry uses these files in an automated fashion to help predict yield problems so they can be corrected prior to the fabrication. Machining, assembly, test and other functions have used these files as a source of information in guiding designers with regard to manufacturability. These files are effective when designs are very similar, but tend to be overrestrictive when using new parts or new technologies, or even when new customer constraints are imposed.

Pattern Nesting. Tools exist which help manufacturing engineers determine the optimal number of patterns to place on a single sheet of material such as sheet metal. These tools help to establish design constraint based upon scrap costs. These tools are heavily used in the printed circuit board industry to influence board design size. Here designs must match material sheets that are limited by tank size.

Fixture Database. This database stores fixture models and information that can be matched with the CAD model. This helps to minimize fixture development. Electronically gripping points, structural stress during processing, setup times and other parameters can be determined.

QC Analysis. Several statistical analysis tools exist to help in planning the proper sampling techniques and the amount of audits, and the required parameters to check within a given process.

Electronic Design Tools

In today's world of electronic design, tool type and usage offer competitive advantages. A typical tool suite for a digital design might consist of:

1) Behavioral Schematic Capture. This is the top level design where engineers can isolate the design into large blocks and only work with the inputs and the outputs of the blocks. This is the level where the design architecture is developed. At this level engineers can view the internal schematics in the box, and make changes if desired. Engineers describe the behavior of the systems independent of any device architecture using high level constructs such as arithmetic expressions.

2) Simulation. Once the engineer is satisfied with the architecture, a behavioral level simulation can be run to validate

Selection Of Tools

the architectural design. Changes can be made to improve the results.

3) Logic Optimization And Synthesis. Once satisfied with the simulation results the logic is optimized and synthesized for device architectures that are chosen later. The engineer can review the design and collapse nodes using high level constructs such as arithmetic operators. This level typically contains both global level and device specified components.

4) Design Partitioning And Automated Device Selection. Once the logic design is established, device selection and design partitioning can begin. This is where the physical design requirements such as power requirements, speed, and package type are considered as well as the list of devices defined by the engineer. These parameters are matched with the behavioral design definition that has been synthesized and optimized. Automated device selection and design partitioning are done. Engineers usually have the options to specify design footprints, partial or complete partitioning, special signal optimizations, and many other details of the physical implementations.

5) Device Libraries. These libraries maintain the schematic and physical details of the devices used. The information may include logic family, speed, power supply current, etc. This information is the source for the automated device selection.

6) Detailed Schematic. Detailed schematic information is generated once the device selections and partitioning are complete. This schematic can be simulated, using device delays. Various combinations of device delays are simulated including all minimum delays and all maximum delays. Newer applications can mix both digital and analog wave forms and the wave forms can be displayed and reviewed by the engineer.

7) Board Layout. The detailed design information is then passed to a board layout tool which places the component on the boards and routes the interconnecting runs. This is a critical area for cost management. Routing tools follow fixed rules; oftentimes minor changes to the board made manually reduce the number of board layers or reduce the total manufacturing costs. Additionally, component placement can be done in a costly manner, so specific information or tools exist to validate that the placement is made in a cost effective fashion.

8) Signal Integrity. Once the layout is complete it is reviewed for potential design problems, such as signals too close, simultaneous switching effects, min-max timing etc. There exist signal integrity tools which check designs for a series of potential errors and highlight those for modification.

9) Manufacturing Process Generation. The layout information is automatically transferred to a series of manufacturing tools to generate the programs for the NC PWB drilling patterns, the mask generation, the component placement programs, and the test generation programs.

Marketing Tools

In addition to engineering tools the marketing and sales functions maintain software tools and databases. These databases consist of customer contact information, such as name, title, function, company, address, previous interests and purchases, and other known data. Thus, the marketing person can identify likely prospects to contact for input, especially if they have purchased a previous product that the new one is replacing. This customer contact information is vital for initial roll-out of products, and in sizing market interest in the planning of products.

Groupware Tools

Information technology provides more than specific application software. The networking of computers and specific tools which utilize this capability are very important in the speeding up of the development process. These tools are discussed below and require that computer systems be networked.

Electronic Mail. This technique has been in existence for many years and allows team members to communicate in an effective way. As questions are identified they can be electronically sent to the person who is responsible for providing the answer. Each person's response is now documented via the electronic message. Therefore, the person responding strives to be accurate and factual, since their input if in error can be traced back to them. Thus, individual accountability allows teams to focus on real issues with factual inputs. Prior to electronic mail, politics played a larger role in problem resolution. This traceability forced many of the development steps and some of the analysis techniques to be more explicit.

In addition, team schedules, action items, customer reports, and test results could be shared in a much larger way.

Notes Files. A second technique that improves the team communication is the concept of notes files. Team members place all the pertinent data in these files or accounts. This data includes assumptions about the design, management directives, test data, contacts, requirements specifications, technical specifications, project schedules, documentation on how various trade-offs were arrived at, QFD's, etc. Anyone on the team has access to this data. Actually anyone with the password into the account can obtain the information. Thus, all team members can see where the product development stands. In addition, team members can use the files for research, leveraging the teams expertise. A team member can ask a series of questions. Others who access the file can provide assistance or a contact. In addition, as new members are added to the development team, notes files provide a way for them to review the history of the project, and understand the constraints that have been placed on the development team.

Product Data Management. Another useful tool is Product Data Management software. This software helps to control the revisions of various files during the development process. Thus, as files are shared, master documents are kept so that only those who have access (granted by password) can go in and change the files. This software facilitates the design documentation process, providing for Engineering Change Notices, documenting the inputs etc. A Product Data Management system provides management of all the product data including test results, and various analysis performed on the product, thus, archiving the data for future reference when changes are made. An effective system provides access from individual workstation to central information databases that are organized into categories by components, assemblies, and products. This data is retrieved for use by the entire team, saving the time associated with recreation. Database search capability also helps designers and manufacturing engineers research similar parts, and similar problems.

Workflow. This is an important tool for development teams. This software allows the routing of messages to a specified set of individuals in a chain fashion, while the originator and others can automatically track the message as it passes through the workflow chain. In addition, work can be routed in parallel. This is especially important when engineers have changes to make. For example, if an electronic designer has questions on manufacturing testing, the designer could route them to the manufacturing engineer responsible, and simultaneously to design engineering test experts to get their opinion. The files will return to the designer with both of their inputs for review.

In another instance, an engineer could work on a specified part of the design, and it could be routed to individual experts for their input. These experts would actually design their section of the product, and it could then be routed to the next person. Documentation could also be developed in this way. Workflow is an important tool for design reviews, sending out electronic files to the appropriate individuals to review either in sequence or in parallel. For instance, a final design might be routed first to the immediate supervisor, then to the design team, without returning to the development engineer and without the need to physically follow up and chase the processing. Workflow used in the review process usually contains automated sign-off capabilities, thus insuring that the appropriate individuals electronically approve the design.

One key feature of workflow is the ability to place deadlines on certain steps. Electronically the team member gets notified that he is late on an activity if the file has not been sent on as planned. This feature facilitates the project management task of finding the status of each and every development activity.

Frameworks. Frameworks allow the integration of various applications. Thus, the moving between applications is made substantially easier. Some frameworks also provide for the integration of the data between tools. Thus, the data itself is converted into a format that can be applied to other tools within the framework.

Importance Of Global Communication. An important attribute of the information architecture is the ability to access a wide range of information. This includes communications with potential suppliers, as well as university research and other data sources such as market research libraries. The products that end up being the most successful will be those that take into account the marketplace change in product offering, technology, customer trends, and cost opportunities. This is information that can be found electronically.

This global information network has already emerged. The effective use of it will depend on the skills of the individual team members, and their knowledge of the existence of data resources and their ability to access appropriate research.

Database Management By Engineers

An important characteristic of a successful engineering team will be a successful database from which engineering information can be accessed and design efficiencies can be gained. In order for this database to exist in an effective fashion it will be necessary for engineers to exercise discipline when using it. If the appropriate data is not stored in the system the database will become useless. The old adage, garbage in equals garbage out applies. It is important for each engineer to understand what information needs to be entered, in what format and which categorizations need to be done, and by whom. For example, a component engineer needs to enter technical specifications, and update them on a frequent basis in order to insure the data is useful to design. Database discipline is essential in a well functioning engineering department.

Achieving Engineering Efficiency Through Application Software

Many software firms will emphasize the design efficiency their particular software will provide. However, the big advance in application tools is the added capabilities, not the productivity. Designing using today's tools provides methods and techniques not even imaginable twenty to thirty years ago. The proper way to gauge the tool efficiencies is to determine if the downstream processes have gotten better. That is, are designs being introduced into manufacturing processes with fewer problems, and fewer engineering changes? Are new technologies being used on a regular basis? Does your product have a competitive advantage? For some companies, upgrading their applications meant that they simply survived the onslaught of tremendous technology change. For others it provided a major advance. The message here is quite simple. True engineering efficiency is measured in successful products.

Putting Together A Tool Strategy That's Right For You

Getting Started. First, when considering a tool implementation strategy the underlying architecture of the hardware must be considered. The software and tool architecture will vary depending on the size of the organization and the type of product. For simple products personal computers may be all that are needed; for others more complex workstations.

The desktop hardware products should be networked together. The networking enables many of the concurrent engineering activities to take place. Networking can be done in a variety of ways. The most efficient and popular one is in a client/server architecture. Here, the desktop hardware represents the client and the system hardware represents the server. Many of the applications such as CAD reside on the client. The server manages the database and provides the network and workgroup tools. A relational database allows engineers to define various search attributes for the products, technical information and specification features.

Networking

Once the server is defined, specific networking protocol needs to be identified and the software products identified. For instance, TCPIP (Transaction Control Processing Internet

Protocol) allows the connection to external networks such as the Internet. Security and control features need to be understood before choosing. In addition to network software, special utilities of interest include load balancing software. This software balances the load among a group of workstations. If you have individually assigned workstations, and at any point in time one or more of the workstations are unused while the engineers using the other workstations find themselves waiting for screen refreshes, then load sharing software may be required to improve productivity. All the software products chosen must operate within the chosen network protocol. Care should be used to select the protocol with the most effective usability in an engineering environment.

In addition, networks can be tuned. That is balancing the server and network activities with those that occur on the desktop. For engineers the objective of tuning is to provide maximum performance at the user level. Thus, network activities that consume the users CPU time or slow down the user's system need to be minimized. To structure a tuned system for CAD applications the desktop hardware will need to have significant power. Since many CAD files are large, usually many megabytes in size, the hardware used will need to be able to operate at high speeds. In addition, memory should be sized so that the files that are being worked on completely reside in memory. The goal should be to avoid disk access during file modifications.

Information Architecture Strategy

Software applications have a broad impact on the development process. Figure 29 highlights only a partial listing of the tools available. With this listing it is easy to understand the problems created if each tool requires recreation of the same data, or is one or two revisions behind in the design, or if the data has to continuously be imported, exported, etc. The implications for engineering productivity and investment dollars are tremendous.

Selection Of Tools

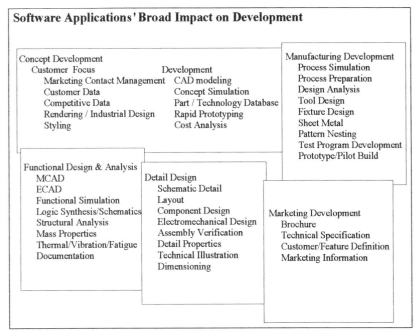

Figure 29 Information Technology Enables Sophisticated Development. *Constant communication is required to maximize collaboration within the concurrent engineering team. As can be seen by this partial listing of CE team applications, information technology is at the heart of this collaboration and is utilized to efficiently develop product knowledge and to efficiently exchange this knowledge between team members. Information technology requires attention to the detail of information flow, data management, productivity enhancements, support applications, and investment spending, in order to rapidly develop products.*

Even the difficulties in priority setting for investment spending or productivity enhancements gets confusing. This is why an information architecture strategy is needed

Selecting specific tools requires a close understanding of what the engineering organization is trying to accomplish. The first step is to define the technology information architecture goal.

Here are a few example goals:

1) Provide CAD/CAE technological advantages for engineers, in order to support the high rate of technology change.
2) Provide simple documentation capabilities.
3) Provide interconnection support for the entire team to facilitate development and shorten both development and manufacturing introduction time.
4) Provide focused analysis activities to eliminate or substantially reduce downstream testing and design validation activities.
5) Develop the design data only once and avoid recreation of data at any of the steps through efficient file management.
6) Maximize flexibility of the architecture to allow different engineers to pick and choose the applications they want to use in the design process.
7) Preselect the tools suite and require engineers and other team members to conform to the standard tools available.

As can be seen each one of these goals is different, and the implementation of a tool strategy will be different. Each of the above seven goals is discussed further. Each number corresponds with the number used for the preceding goal statements:

1) A CAD/CAE technology based strategy, that is, one that chases the advances in technology to provide world class capabilities, is an expensive approach and should only be used for those high technology products that need this level of functionality and can adjust to the constant changing.
2) A simple documentation strategy is the least expensive. Good documentation packages can be found as personal computer products.
3) Providing interconnection for the entire team requires database management and groupware focus that supports the CAD and CAE applications that exist.

Selection Of Tools

4) Providing focused analysis activities requires research and development of certain areas, and oftentimes modifications to tools that exist in order to achieve the results needed. Since substantial simulation is required this sets a direction regarding hardware selection and performance distribution among the design activities. This is by far the most powerful of all the strategies since it eliminates many test operations after the design is complete. Since testing of the software and validating the analysis against the downstream product testing is required, this strategy requires more than a simple software and hardware implementation.

5) Developing design data only once requires a focus on frameworks and data translations among tools or the selection of a single company's products. This data architecture is good for pointed productivity needs, but is often inflexible in achieving the full spectrum of engineering requirements.

6) Providing a flexible architecture will allow engineers to pick various applications but requires them to translate the data or recreate the information needed. This strategy is excellent at providing a broad spectrum of tools and applications for the engineers and in staying current with the latest technologies available. This must be traded off against the constant data recreation problem.

7) Preselecting tools is a typical top down approach to engineering. It provides no flexibility for engineers to peruse other tools. It does provide focus on the tools at hand to improve and integrate them, or modify them as appropriate. A case can always be made for a single integrated tool suite. This, though, puts the engineering department at the mercy of the software supplier for maintaining the latest in technology at a price the company can afford. This strategy has been found to be risky in the long term.

Thus, choosing the right information architecture strategy is very dependent on the goals you intend to achieve, the risks you want to take, and how to position the needs for your products against the investment needs of the engineering process. For many companies, using a simple documentation strategy coupled with an excellent communication strategy will provide an answer to product design. Other companies will recognize the potential to dramatically change the development process as a result of implementing information technology and the power it brings. Those companies that use these strategies as the foundation for adding new engineering capabilities, improving communications, redefining the roles of the development team, and improving engineering productivity will win in the long run.

Client /Server CAD

One of the most efficient information architectures that is in use today is client/server CAD. This is a foundation technology infrastructure which facilitates rapid product development (see Figure 30). It provides the capability to effectively use multiple CAD/CAE tools and applications packages by the full spectrum of the development team, including analysis groups, testing groups, reliability, quality, manufacturing, and even supplier and customers.

Client/server CAD applications are distributed throughout the networks. Individual workstations or PCs (the clients) handle the applications and the data files reside on the server (see Figure 31). For example, CAD runs on a workstation or PC client. The CAD tool fetches the file from the server and loads it into a temporary storage area on the client. Periodically, and when a task is complete it writes the data back to the server where it is stored in the database. There are significant advantages in security, revision control, data management, and backup protection, since the server can be centrally managed. Additionally, even if the file is checked out by one client it is still available for viewing by other clients.

Selection Of Tools

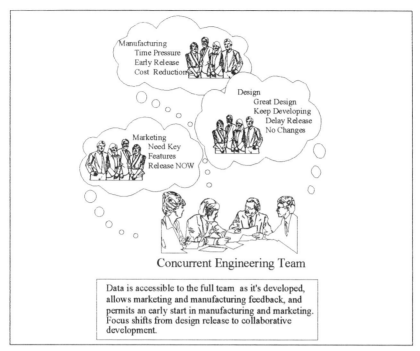

Figure 30 Client/Server CAD Maximizes Collaboration. *Client/server CAD is a narrow definition of the industry understanding of client/server. The key is to maximize collaboration by providing data access to all team members. In addition, this allows user flexibility in their application selection, while centrally managing product data, security, back-up, etc. Work can easily be distributed among engineers. Client/server CAD matches the information infrastructure to the Concurrent Engineering Team.*
(The artwork is derived from Lotus SmartPics for Windows. ©1991 Lotus Development Corporation. Lotus and SmartPics are registered trademarks of Lotus Development Corporation)

In client/server CAD the data management problem is solved. A single database avoids confusion in the development process and with product data management software eliminates the

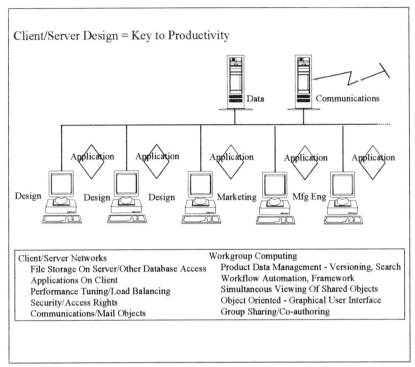

Figure 31 Client/Server CAD - Focused Definition.
Client/Server CAD improves productivity as well as speeding up the development cycle. It couples effective information technology design, engineering application software, and workgroup computing tools to simplify many of the project tasks.
(The artwork is derived from Lotus SmartPics for Windows and Lotus Freelance Graphics. ©1991 Lotus Development Corporation. Lotus and SmartPics are registered trademarks of Lotus Development Corporation)

problem of one engineer working on the wrong revision of the part. Tools exist to make multiple databases appear as a single database to the client. Client/server data file management is a significantly more efficient method of distributing work throughout the development team. It facilitates multiple development of products and sub-assemblies. Parts can easily be

Selection Of Tools 175

distributed to multiple engineers for development, and they can validate how their piece fits into the entire system using the latest revision of the design. Manufacturing can review the pieces as they are being developed and provide suggestions on improving the manufacturing costs.

Utilizing workgroup tools with the client/server CAD concept allows for an engineering repository automatically stored as data in a relational database. An object oriented desktop makes it simple to access, share, and communicate through the entire organizations. Workgroup tools eliminate repetitive tasks such as those associated with file transfer. They can be set up to provide document management, workflow, ECO processing, drawing approval, work schedules, and other productivity improvements needed for efficient engineering.

Tool Limitations

In thinking about software tools, it is also important to understand their limitations. In today's environment, tools are big business, and because of this they are oversold in capability and ease of use. The very fact that experts are needed to operate key analysis tools causes dependencies to be added to an already overcrowded development schedule. Data translation from tool to tool introduces difficulties which are not normally found in the skill set of the average engineer or team member, and introduces a database management revision control problem. For example, adding a simple access hole may require changes to several different databases. Different tools are called into play, including MCAD, Finite Element Analysis, sheet metal process tools, assembly design analysis tools, and data management. Although valuable insight may be gained from this process, the cost of using this process must be considered. It is not for everyone.

It is clear that database change management must be planned, or expensive support activities will result. Every change cannot be verified as it's defined, but rather groupings of changes are validated. For low technology products the economics of tools and their use are very questionable: Does the cost of the tools and development of the tool's experts and data management provide the return needed to justify their existence? For more sophisticated products the question is still a valid one, but is more formulated towards capability, i.e., Does the capability provide added value?

Additionally, once the tools are established, the quality of engineering shifts to the "tool users". Thus, drawing detail is the CAD person's role, structural integrity becomes the FEA expert's role, etc. All come under the direction of the project engineer. This shift of roles and responsibilities as well as organizational limitations is one that must be address for effective CE. For cost reasons one would suspect that locating the analysis activities in a central group to be parceled out as needed is the most cost effective. However, the autonomy of an indirect engineering member makes the project more difficult to control by the project engineer, and the less useful the CE concept becomes. Again, tools affect roles and responsibilities and underscore the importance of a thorough understanding of the CE concepts as they are implemented.

Summary

This chapter identified some of the tools that exist for engineers in the development process. Each tool was described and positioned as to its usefulness in the design process. No specific software vendors tools were described in detail nor were the tool's usage described in detail. The purpose of the information was to provide an overview of the software technologies, focusing on some of the more popular ones. They are listed below for quick reference. In addition, the information architecture strategy and client/server CAD were discussed, as well as the limitation of tools, and an insight in proper implementation was suggested.

Selection Of Tools

Mechanical Development Tools--MCAD/MCAE
    2D CAD
    3D CAD
    Surfacing
    Solids Modeling
        Types of Solids Modelers
            Explicit Modeling
            Parametric Modeling
            Variational Modeling
            Feature Based Design
    Detailing
    Finite Element Analysis
    Dynamic Analysis
    Rendering Tools
    Information Libraries
    Spreadsheet
    Thermal Analysis
    Design Optimization

Manufacturing Engineering--CAM, CAE
    Electronic Assembly
    Tolerance Analysis
    Mold Flow Simulation
    Assembly Analysis
    NC/CNC Programming and Simulation
    Sheetmetal Unbending
    Rapid Prototyping
    Process Simulation
    Manufacturing Process Files
    Pattern Nesting
    Fixture Database
    QC Analysis
    Test Pattern Generation

Electronic Design--ECAD, CAE
    Behavioral Schematic Capture
    Simulation
    Logic Optimization and Synthesis
    Design Partitioning and Automated Device Selection
    Device Libraries
    Detailed Schematic Capture
    Board Layout
    Signal Integrity
    Manufacturing Process Generation

Generic Tools / Databases
    Information Library
    Parts Libraries
    Networking

Marketing Tools
    Contact Automation

Groupware
    Electronic Mail
    Notes Files
    Product Data Management
    Workflow
    Frameworks

Global Communication

Client / Server CAD

Chapter 6
Market Focus Your Design

What Is Understanding The Market?

What does it mean to understand the market? This seemingly undefinable phrase can have very specific data and information. One can define this as understanding what it takes to satisfy customer needs and to reach those customers. This understanding can be quantified in specific features required, price customers are willing to pay, positioning the product against the competition, barriers to sales, and understanding the distribution channels required to reach the customers targeted. In addition, there are geographical understandings, and regional understandings of the market that are often important to act upon. This chapter addresses understanding the market and translating that into product needs, and the team activities needed to insure success.

Selecting Customers Who Count

One of the most common mistakes that is made in market research by development teams is the lack of market testing with the right customer base. The right customer base consists of those customers who are expected to either pay for or use the product. In most cases this is more than one person. For example, suppose you designed testers for use by telephone repair persons. The repair person has certain requirements regarding usability, accuracy, weight, etc. This person's requirements can be considered user input. But more than likely only the purchasing department and the manager of the repair function are going to determine what gets bought. They are going to look at features and prices and determine which features are worth paying for. If research is done with the repair person, they will have difficulty accurately separating the features they want from the features they need and will give the impression that more features are better. This is not the case. Cost effective features are better. Thus, the person who must select and pay for the device is the decision maker who should be the right customer to target, even though the user may be different.

A second targeted customer can be found by identifying those customers who would use your product in volume and solicit their input. Take for example videotapes. A large purchaser such as Blockbuster Video can provide excellent customer input since they have such a wide usage of this product.

Here are a few guidelines to follow when identifying customers who count:

1) Select managers who purchase the types of equipment you sell and secondarily those who would recommend the selection of products.

2) User inputs can help identify features that are useful but only the purchaser or recommender can distinguish between the features and identify value.

3) Look for volume purchasers of the products, since their experience will be broad based.

4) Look for customers who have recently purchased a similar product. They are likely to be more realistic around affordable features and options.

Getting Good Customer Input

Once the right customers to target for input are identified, getting them to provide input is another question. These people are usually very busy, difficult to reach, and when reached often non-committal. To overcome these obstacles and others, there are some well tested methods. First, companies establish customer review boards. The boards are usually comprised of senior managers from the company's major customers. For the customer this provides a way of keeping up with the latest product offerings and in influencing their development. For the company it identifies a valuable contact for product information and general business and customer satisfaction issues.

However, working with review boards is not always an option to product teams. A second way is through independent market research. This can take many forms. Telemarketing is one of the methods. It allows a few questions to be answered and can take place rapidly and over a broad spectrum of companies. It is limited by the fact that few people will be willing to go into any real depth over the phone.

Another method of market research is individual customer interviews. This usually takes place with a researcher and a member of the development team. These people travel to a customer's site and interview the customer for a hour to obtain their direct inputs on your ideas. This allows more subjective input and a much better understanding of the customer feedback. In addition, visual props such as prototype products can be used. It also allows the engineers to ask a few key questions. The danger in this research is that it is limited to the few individuals who are contacted and may not represent the market.

Another method is called focus groups. Focus groups are useful where there are large concentrations of decision makers. Here the customer is asked to attend a market research activity with his fellow peers from other companies. The session can be blind, that is, the customer does not know who is sponsoring the research. They are asked various questions, shown alternatives, and asked for their candid inputs. The session can be watched through a one way mirror by the development team while in progress, since it is run by a research expert. The disadvantage of the focus group is that individual input can often be overshadowed by group input, and their responses influenced by group attitude good or bad. One strong leader in the group can sway all their opinions and give a distorted market reading.

Still another method is to send potential customers either product samples or questionnaires, and pay them an honorarium for providing the input data.

As one can see, getting good data is difficult, with the potential of being inaccurate or misinterpreted. Since this is the case and the data is the basis of the product features selection, it is important to recognize the risk to the development team. Market research data is limited in scope. The solution is to provide multiple market research efforts occurring at appropriate times during the product development cycle. Consistent marketplace messages and themes can only be found over a period of time. This allows the customer input to become more substantive. It also allows the team to check on market changes and competitive product changes that occur.

Focusing The Data To Reflect The Market

When market research data is obtained, the source of the information is a key input. Does this come from individuals who represent the paying customer? Does it come from the companies in the mainstream of the market you are targeting? For example, if your product is for electronic manufacturers, and the predominant feedback you received is from semiconductor manufacturers (a segment of the electronic market), then you may be basing your product decisions on too narrow a band of companies.

Market Focus Your Design 183

Furthermore, if you are targeting engineering managers as your key paying customers and the data is from a mixture of engineering and manufacturing managers, then some data is more useful than others.

Another major consideration is region. One can design a product for many markets, but if there is no way to effectively get the message and product into these markets, then the work is wasted. Thus, it is important to understand the regions where distribution exists so that the work can be tailored to specific market targets within that region.

Thus, it is important to sort the market research data by:

1) Market Segment (example, manufacturing industries)
2) Market Sub-segement (example, electronics)
3) Target Customer (example, V.P. of Engineering)
4) The person (by name) and the company (by name)
5) Geographical region (sorted for distribution channels)

Understanding Market Penetration Through Product Design

In order to understand market penetration, one must first understand market segmentation. Market segmentation is a key activity that allows the correct interpretation of the market potential. It will also allow you to determine which market segments and which customers can be the prime targets for your product. For example, markets can be segmented by product type using SIC codes. Each company in the United States has an SIC code, and databases exist which list companies by SIC. An example SIC code listing is shown in Figure 32.

Thus, one could find out how many companies in the United States produce furniture, or how many produce medical devices. Once the market segmentation is complete, the number of companies by market region can be identified. Those regions

184 Chapter 6

which have large concentrations of target customers ideally should be the ones where your distribution channels are the strongest. An example regional description is shown in Figure 33.

Number of Companies by Company Size (In Millions of Dollars)							
SIC	Description	≤5	6-25	26-50	51-100	101-350	>350
5047	Medical, Hospital Equipment	5	1	1	0	0	0
5045	Computer, Peripherals & Software	12	12	0	0	1	2
3544	Special Dies and Tools	20	1	0	0	0	0
5511	Car Dealers New & Used	25	54	4	1	0	0

Figure 32 SIC Code Examples. *All companies are required to be identified by SIC code. This code is usually by site and focuses on the number of employees at that site. Several databases use this information to project sales volume. This information is the basis for targeting specific markets*

SIC	Description	SIC	Description
2511	Wood Household Furniture	3563	Air and Gas Compressors
2522	Metal Office Furniture	3568	Power Transmission Equipment
2542	Metal Partitions and Fixtures	3572	Typewriters
3465	Automotive Stampings	3639	Household Appliances
3531	Construction Machinery	3641	Electric Lamps
3536	Hoists, Cranes and Monorails	3661	Telephone and Telegraph Apparatus
3541	Machine Tools, Metal Cutting	3675	Electronic Capacitors
3555	Printing Trades Machinery	3713	Truck and Bus Bodies

Figure 33 Regional Market Information. *This can get very specific for targeting industries, companies, and eventually individuals. This information is the basis for leads being generated for sales. The above information is representative of market information by state county.*

Market Focus Your Design

The next piece of data is more difficult to get; that is, to understand by key region which percentage of the market your products have. From standard databases the number of potential customer companies is known. Now this needs to be converted into market share information and market readiness information.

The method to use in this information gathering session is telemarketing. Many key customers can be called and asked three key questions:

1) Which product do you own? (This provides market share).
2) Are you anticipating any purchases in the near future? (This provides market readiness/ short term opportunity).
3) If so, which product would you buy? (market shifting).

The data can then be summarized and you will know several things:

1) The percentage of market your products own.
2) Your top two or three competitors by region.
3) The percentage of customers looking to purchase new products in the near future indicating the rate of growth or shrinkage of the market in the specific region.
4) Customer satisfaction with their current products. In other words, how many customers are looking to change from their current products.

The last step is to understand the market potential of the product you are working on. Assuming that through Quality Function Deployment techniques the product team understands the appropriate features for the product to be competitive, the next step is to understand if customers are likely to purchase these products. To do this the following steps are used:

1) Identify the market share your company currently has and of that how many customers are looking to purchase additional products in the near term. If they have your product and are looking to purchase additional products, and your new product

is the cost/performance leader, then there's a good chance that a high percentage of the market will be yours. The percentage can be confirmed through telemarketing.

2) Identify the market share owned by the competition and of that number which percentage is looking to purchase new products in the near term. This is the market growth potential. However, there are many non-product barriers to penetrating this market. For example, the customer's desire for product consistency, or the competitor's product has been customized in their unique environment, or a specific sales relationship exists. Focused groups using prototype products can provide some understanding of these barriers. In many cases the customers will identify specific reasons for not purchasing your products which are non-product related. From this data a subjective indication on market penetration can be postulated. For example, if twenty out of a hundred who are ready to purchase and have no barriers are thrilled with the new product, then the likelihood of a 20% penetration over time is good. However, any that are less than thrilled or have identified barriers to purchasing indicates that non-penetration is likely no matter how good the price/performance metric. Note: barrier reduction or elimination (usually by sales) is needed before sales can be considered likely. For major projects, the data from several focus groups in different regions should be used.

Marketplace Competitive Analysis For Your Design

Once a top product is developed, getting the market to understand the competitiveness of the new design is difficult to do. Much rests on the capabilities of the market organization which support the new product. Competitive products need to be benchmarked tested, reported in the press, advertised, and demonstrated to customers before they can be recognized by the purchasing customer as a better performing product. If the marketing organization has a strong relationship with the test agencies, the press, and the trade journals of the specific industry, then there is a good chance of success. If these relationships are not strong, the data is likely to be viewed skeptically by the press and customers and the chances of establishing the correct product perceptions are reduced. If, in addition, no testing house exists that routinely runs compassion testing, then the product's competitive advantages are going to be very difficult to establish. Incidentally, much of this credibility must be established by market and by region.

A small focus group can be used to determine how easy it will be for the market to understand the competitive and feature advantages of your product. In addition, if the product advantages will be either expensive or risky to establish, then the market potential of the product will be diminished. On the other hand a strong marketing organization can help a weaker product meet its target by developing the right image in the press.

Determining The Correct Market Positioning For Your Design

Positioning the product correctly in the market is a function of features selected, price, market reputation and advertising. One advantage the larger companies have in launching a new product is market presence and reputation. This generally allows their products to be priced higher, since customers want to purchase from dependable companies who can back up their products.

The second thing to examine is the newness of the company's products to the specific market. If the product is new to the market then the positioning of the product will need to be more aggressive.

Third, set a price for the product that is consistent with the functionality. The price can be slightly higher if the companies reputation is good, and the price should be decreased if it is the first company offering in a new market. For unknown company's aggressive pricing is needed in order to get established.

Product price positioning can be developed by comparing functionality to the existing competition. The features and the pricing should be consistent between products. Thus, more features should be priced higher, and fewer features should be priced lower. In addition, pricing should be set based upon growth targets and company reputation in the marketplace. Price the product lower than the norm for aggressive growth, and higher than the norm if a good reputation is established. However, if the market is flooded with products, then reputation plays a larger role in setting a price. It is important to remember that low price in the customer's mind is not always seen as advantageous. Customers often equate goodness with price. Thus, a new product can be priced higher if there are some discerning characteristics that add value from the customer's viewpoint.

Examples Of Good Market Positioning

Let's examine some examples of effective market positioning. The first example is a product from Chrysler called the Viper. This car was developed using a concurrent engineering approach. The Viper is Chrysler's high priced sports car and it provided a challenge for them. It was their first product in this high priced competitive sports car market. To meet this challenge Chrysler needed to position its product properly in the market. The first step was feature definition. They defined their market features as the beauty and performance of the cars of the 60's, and to be the successor to the Shelby Cobra. This meant power and excellence in design and excellence in driving. With regard to power the Viper can go from 0-100 mph and back in 14.5 seconds. A sports car enthusiast's dream. Additionally, they added a new technology that reduced weight and added beauty. They elected to use

Market Focus Your Design 189

Resin-Transfer Molded parts which allowed sporty contours and a highly attractive design.

Once the key features were established the price needed to be set. The $50,000+ price suggests the low end of this class of car. Despite their excellent reputation as an automobile manufacturer, they still needed this low price to attract the market since they did not offer products targeted for this class of customer.

Next, Chrysler needed to identify its advertising plans. Since their market was sports car enthusiasts the way to target this market was through sports cars shows and racing events. The Viper won the 1989 North American Auto Show as a concept car. In the spring of 1991 Chrysler made a decision to use their newest prototype as the Indianapolis 500 pace car. With the attention it received from the drivers, the press and those in attendance, this turned out to be an excellent decision regarding the launching of this product and clearly positioned it as a sports car enthusiast's dream car.

Thus, Chrysler selected features that would win in the marketplace, and they priced their car recognizing that this was a new market for their company. They used their reputation to help promote the car (Auto Shows), and they launched the product with all the attention possible in this niche marketplace. Thus, it was no surprise when they were backlogged over 2 years shortly after the product started shipping.

A second example is one involving Craftsman Tools. In this case feature differentiation was not always obvious to the customer. The general perception was a hammer is basically a hammer. Therefore, the lowest priced hammer is the best one to by. Craftsman took a different tack. They decided to produce and sell a high quality line of tools. Their market positioning was their guarantee, "if it breaks we replace it"--a lifetime guarantee. This market positioning allowed them to charge a price over 50% greater the those companies chasing the cost-sensitive market positioning. Since customers equate pricing with quality, the price actually helped to sell more products and promote the image of high quality tools. This positioning was intended to attract the construction trade experts, automechanic experts, the home market, and others who use hand tools. The name Craftsman suggests their product image, tools used by the most talented and skilled workers. Another excellent example of effective market positioning.

Summary

This chapter was a brief outline describing the basic methods to use in understanding the market potential of your unique product. It described how to search out the right customers, the type of data to expect, and a simple market research method of understanding the potential for your product. It identified the data and method to use regarding market penetration, pricing positioning and suggested when to use aggressive marketing tactics. Lastly it identified examples of good market positioning.

Chapter 7
Developing Cost Sensitive Products

Understanding Product Costs

Understanding the basics of product cost starts with understanding two key product development areas; selection and design. Selection includes those items that are selected for use in the product, such as components, materials and sub-assemblies. Some of the selections are obvious, others are more subtle. The second key area is the design activity itself. That is, the choices made by the design engineer during the development process, such as geometric shape. Like selection some of these choices have obvious implications for product manufacturing costs and others are not so obvious.

In order to understand product cost the engineer must examine those items that are defined by the requirements as selection and those that are allowed to be designed. An example of this is a piece of test equipment with a small 2 inch wide printer. The printer is likely to be an item of selection whereas the enclosure and the electronics are likely to be an item of design. This is an important distinction since the techniques used to minimize costs are significantly different.

In addition, the area where these two concepts intersect is usually a prime target to minimize costs. For instance, the printer can be purchased with an enclosure of its own, with a bezel, or even with special cables, or more simply, the print engine can be purchased and the control electronics designed by the engineering team.

Determining the level on which a design team should attack the cost problem depends on the skills of the team and the types of risks the engineering team is willing to accept, as well as the product requirements.

A strong example of the importance of selection can be illustrated with an electronic component. The selection of an electronic component affects the electrical design and it affects the component placement on a printed wiring board and the printed wiring board design itself. It affects the test strategy for the component and the automated assembly process. Thus, with a single component choice much of the associated cost becomes set. It is important to have strategies and methods to address cost reduction at the time of selection.

Engineering Selection

Feature Selection. The first area that is meaningful to consider is the actual features selected. In general *the fewer the features the less costly a product will be.* It is important to evaluate each feature by its ability to provide marketplace advantage for the product. This must be a careful trade-off since some features will turn out to be "customer delighters" and provide a strong reputation for the product, while others will simply be a wasted effort. It is very difficult to tell which features

Developing Cost Sensitive Products

provide the biggest market return at the onset of the development process. Thus, many of the techniques mentioned in previous chapters, such as Market Research, Customer Focused Design, Quality Functional Deployment and Pugh Analysis provide useful ways to understand and evaluate these critical features. The trick then is to minimize the number of features chosen.

Options Selection. The second area of consideration is the selection of product options. Options in this usage refer to customer selectable items. For example, with a personal computer options would consist of display panels, disk drives, keyboards, etc. For the automotive industry, this might be radio selection, phone compartment, electronics controls, power windows, etc. These options are then configured into the product and shipped as a completed assembly.

The way to minimize option costs is to select standard high volume options and avoid unique requirements or specifying changes. High volume options where practical will provide both lower cost and product dependability. In some cases in may be practical to purchase high volume options and modify them on site to meet the needs of the product. This may even mean some disassembly of the volume option.

If the high volume option alternative is not available, the next choice level is to look for options or components which meet industry standards. This allows for a more competitive positioning when purchasing is contracting for volume product. Standards such as interface requirements, and bus structures all insure competitive quality. Suppliers are striving to provide standard parts at the lowest costs. Thus, using these standards allows the designer to take advantage of the investments made by the suppliers who provide these products.

In addition, the purchasing department can only maintain a strategic focus with a limited number of vendors. With these vendors the purchasing department can work special pricing arrangements. In addition, once these vendors are selected they invest in themselves to better support your product needs.

Typically, too few engineers manage to take advantage of both industry standard components or preferred vendors, and therefore don't leverage these cost reduction investments into their designs.

For components, many companies develop preferred parts listing. These listings identify components with excellent quality and delivery histories. They also identify technical specifications and typically they identify specific vendor processes that have been certified to insure consistent quality.

Sub-functional Analysis. The next choice in the selection process is to determine the minimum level of assembly in which the options can be effectively incorporated into the product. The goal is to eliminate a substantial portion of the costs and product mark-up of the purchased part. For instance, it may be more effective to design a 2-inch wide printer into the product using only the printer mechanisms from the supplier. Thus, one must look at each option and functionally subdivide it to determine if the sub-functions can be designed by the team with positive cost results. A common methodology used here is a process of branching to identify cost targets and matching those with team skills. This process is shown in Figure 34.

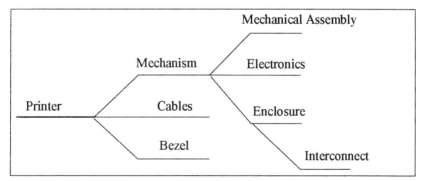

Figure 34 Branching Methodology. *This helps to isolate those parts that are unnecessary for your application, or can be designed to be less expensive.*

Developing Cost Sensitive Products

Team With Suppliers. When high volume options are not available, or your purchase quantity is a high volume by itself then *teaming with the key suppliers usually results in the lowest costs both short and long term.* The supplier should determine technologies that are emerging that will reduce the cost of the options or parts. The supplier may have ideas not yet considered by the design team, or he can identify products that are emerging that will provide a cost advantage. This is true in the area of materials such as plastics, or in the area of components, or in the area of sub-assemblies. Suppliers who know their business and have agreed to work with you in the development of the product can be a valuable source of product cost savings ideas. In addition this long term partnering often results in the lowering of mark-ups. Teaming with suppliers allows the designer to influence the suppliers own investment for both short term and long term gain.

The Design Of The Interface. Where the option and the system interface with one another, a number of selections exist mechanically and electronically. These need to be reviewed carefully for savings, since saving at the purchase option level usually means the addition of assembly and testing internally. *Experience often shows that the minimum level of purchase is the lowest cost for volume products and the highest cost for low volume products.* This is because the design effort cannot get amortized over the volume of the product in the low volume situation. For example, a product with a lifetime volume of 500 pieces must have a savings of $200 before a one person-year effort can be recovered (assuming $100,000 is the fully burdened cost of an engineer).

In addition to the option cost savings, internal cost savings can occur based upon the design of the option installation process and the test time required. Thus, the selection of the interface is closely tied to the system manufacturing costs. Since option slots can usually accommodate multiple options the cost optimization problem looks like the chart in Figure 35.

	Annual Installation Costs	Annual Option Volume	Unit Installation Costs	Interface Parts Costs	Installation Assembly & Test
Option A	$ 330,000	11,000	$30	$26	$4
Option B	$ 224,000	8,000	$28	$19	$9
Option X	$ 366,000	6,000	$61	$45	$16
Left Empty	$ 240,000	12,000	$20	$12	$8
Total	$ 1,160,000	37,000			
Cost Per System	$ 31.35				

Figure 35 Option Cost Analysis Table. *For product options the total cost cannot be calculated by a simple Bill-of-Materials costing technique, since these parts are not used on every product. Their cost must be adjusted based upon volume, as shown in the table above. The total annual installation cost is a summation of the products of the annual option volume times the option (unit) installation costs. This total can then be divided by the annual system volume to give the average installation cost per system for a single option slot. As can be seen, even though Option X is the lowest volume it has the largest potential for cost reduction.*

By looking at the total cost of the installation the proper cost reduction decisions can be made and the true cost of the option selection can be addressed.

This formula can be used for each option slot to determine where the real cost problems are with option installations, and to prioritize the design efforts.

Developing Cost Sensitive Products

Make Versus Buy Decision. Make versus buy is a typical decision that leans heavily towards buy when cost is the primary determining factor. It is difficult to compete on a price basis with companies that are very product focused. For example, power supply vendors are very numerous. The design effort and engineering skills for cost effective power supplies are very focused. Even though these people could be hired by larger companies, more and more often these companies are finding that internal groups cannot match the cost effective solutions offered by power supply vendors. In addition, many of these focused vendors make huge investments in improving specific components or sub-assemblies, or in developing competitive processes. The engineer can take advantage of these investments by seeking out these suppliers.

Cost Management In Design

There exist several basic techniques for understanding product cost during the development of the design. These techniques can be used when the design is in process so that the inputs can be incorporated into the product. Good cost management of the design begins with organization, and utilizes the full development team's knowledge. The first step in cost organization is to develop a parts list and cost strategies for minimizing the costs. Even though the real design isn't done, a "concept" parts list can be developed based on previous products. This list should contain all anticipated parts, plus those than can be assumed to be similar based upon a close examination of the previous product. Once a representative parts list is established, then cost targets and strategies can be developed. These strategies might include conversion of parts to plastics, or the development of an ASIC device to replace a series of electronic components. Many of these strategies have been previously covered in Chapter

4 under the concept of Design For Manufacturing and Cost Driven Design.

Parts lists and parts costing can be done in many different ways; some of them hide the true costs of the product. For cost analysis the part and cost lists must represent the complete product cost model for the entire product, and it needs to be tabulated in a way that allows the entire product development team the ability to make trade-off decisions. An example of the categories for tabulation are:

1) Cost of material by part numbers
2) Cost of sub-assemblies by part number
3) Cost of customer selectable options by option number
4) Cost of assembly in units of time
5) Cost of test and repair in time and repair (or scrap) dollars
6) Cost of inspection time
7) Cost of shipment packaging and associated labor
8) Cost of selectable options installation costs (parts & labor) by option
9) Average cost of warranty service parts, labor and travel

The complete cost model can then be calculated and combined with the development budget to make effective cost trade-offs. Thus, using the numbers from above the following cost table can be developed as a tool of the design team. All final cost calculations are on an annual basis. Volume numbers should be for the second year of production volume to avoid the ramp-up calculations. Each calculation is for an individual item, and the summation of all individual items is equal to the total for that item. Thus, the parts total is a summation of all the annual parts costs (item #1). The calculation for costs can be shown below:

Developing Cost Sensitive Products 199

Annual Part Costs = Σ_1^n (cost of material part number N) x (part quantity for N) x (annual product volume)

Sub-assembly Costs = Σ_1^n (cost of sub-assembly N) x (sub-assembly quantity N) x (annual product volume)

Option Costs = Σ_1^n (cost of option N) x (percent of option usage for N) x (annual product volume)

Assembly Time = Σ_1^n (hourly cost of assembly N) x (hours for assembly N) x (annual product volume)

Test and Repair Costs = Σ_1^n (test and repair cost per assembly N) x (quantity of N assemblies) x (annual product volume)

Inspection Costs = Σ_1^n (cost of inspection per part or assembly N) x (quantity of N part or assembly) x (annual product volume)

Shipment Materials Costs = Σ_1^n (shipment material costs N) x (annual product volume)

Option Installation Costs = Σ_1^n (cost of installing option N) x (percent of option N usage) x (annual product volume)

Warranty Costs = Σ_1^n (anticipated number of service calls per component N per year) x (cost of service call for part N)

Once the annual cost table is developed, benchmarking can be used to insure competitiveness. For each of the items 1 through 9 a competitive cost factor can be determined by tearing down a competitors product and costing it out as if the host company produced it in volume. This benchmark will also drive strategic thinking regarding costs. This method is part of the Cost Driven Design approach. Thus, the final cost table is show in Figure 36. Finally, a total cost is derived and each item is calculated as a percentage of the total and entered into the chart.

Cost Target	New Design Costs	Targeted Costs	Previous Product	Competitive Benchmark	Engineering Budget
Cost of Material & Part A	10%	8%	20%	20%	30%
Cost of Material & Part B	20%	20%	30%	10%	20%
Cost of Sub-assemblies	40%	32%	20%	30%	24%
Cost of Options	12%	20%	8%	25%	9%
Cost of Assembly (time)	2%	4%	6%	1%	15%
Cost of Test & Repair	2%	4%	2%	1%	1%
Cost of Inspection	2%	4%	2%	1%	0
Cost of Shipment Mtl + Labor	4%	3%	4%	4%	0
Cost of Option Installation	3%	3%	3%	3%	1%
Cost of Warranty	5%	2%	5%	5%	0

Figure 36 Cost Driven Design--Cost Trade-Off Analysis. *The above table shows the trade-off matrix that can be used to compare the specific product cost targets to the engineering budgeted expenditures to be spent during development. It will help to identify if sufficient spending is being placed in the right areas to drive costs to hit their target. The engineering budget column identifies the percentage of budgeted dollars to address a specific area of cost. Typically the greater the delta between the targeted costs and the previous product the greater the engineering budget should be.*

Design For Manufacturing

There are several basic areas under the concept of Design for Manufacturing. These areas include:

Developing Cost Sensitive Products

Design analysis
Process analysis and flow
Design for assembly approach
Design rules
Technology analysis
Tolerance
Others

Design Analysis. One of the basic premises of design is good analysis. For example, good structural analysis will help to determine the adequacy of the gauge selected in a mechanical enclosure. It will also help to minimize the supporting structure needed. Thus, standard design analysis done well can minimize structural requirements and material costs.

A similar analysis can be done in the electronic design. Signal integrity simulation can provide the analysis necessary to validate that errors due to timing and due to signal degradation can be avoided. Thus a good analysis can minimize the part content from the outset of the design effort. Analysis may seem an obvious area, but in practice design engineers are reluctant to spend the time or money to do the analysis for cost purposes, once they know the design will work.

Process Analysis. In order to minimize the cost of a single part the process steps must be minimized. Thus, by performing a detailed process analysis the cost determining factors can be identified and minimized. For instance, in enclosure design the process could include bending, punching, welding, sanding, masking and painting. In this process standard size sheets are used. If the parts designed can be sized to fit in multiples on the sheet such that the excess material will be minimized, then the part costs can be minimized. In addition, if the process steps needed to complete the design are such that they are in the wrong sequence for the most efficient manufacture of the part, unnecessary costs will be incurred. This may happen if all the punching can't be done on one trip to the press due to setup or directional limitations, or if multiple weld steps need to occur both before punching and after punching.

The best method to determine the optimum manufacturing processes is the flow chart. The flow chart method shows the process steps as designed and the costs of each step. It identifies where costs must be removed to minimize the costs of the product. In one such analysis the masking and hand taping of a side skin represented the majority of the paint costs of the part. A simple change in the edge design minimized the masking labor content. This is one such example that can be attributed to the effective use of flow charting prior to design release.

The flow chart method of analysis can be seen in Figure 37. This method shows the areas of duplication, and also shows those areas of costs which add no value to the product such as inspection and testing. Thus, both the designed in costs can be identified and the quality controlled in costs can be identified.

Concept And Cost Reviews. Since the majority of the costs are locked in at the concept phase it is important to influence the design decisions early in the design process. The best method of doing this is to review the design ideas with key technology experts. These experts can be found in many places, including manufacturing and suppliers. These concept reviews are not intended to critique the product or determine its readiness, but should be very useful to the designer in soliciting the opinions from experts and enlisting their help. A good concept review will provide the input needed to pursue a different path in the development process. Thus, formal concept input reviews can provide key ideas to pursue and review, and can allow the usage of an expertise upon which the designer can rely when developing the product itself. Many a good idea is lost because the designers are not comfortable implementing the suggestion with their unique skill set. A good review with support experts should avoid this problem.

Developing Cost Sensitive Products 203

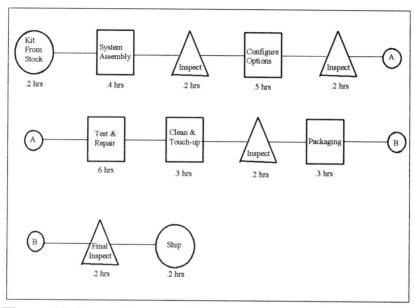

Figure 37 Flow Chart Analysis--System Assembly. *This is a good way to understand manufacturing costs and target cost reductions. In the case above considerable time is spent on inspection, and could be minimized. The first inspection step and possibly the second one can be eliminated if the assembly effort and configuration was made simpler and less prone to error.*

In these early reviews many alternatives and various options regarding costs should be explored. Electronic modeling can be used to show the effect of various design decisions on the product. It is important that those people who understand the detailed manufacturing processes review the concepts and make recommendations. These process experts (typically producibility experts of manufacturing engineers) can highlight areas of potential savings and help to developed the appropriate cost targets for the product. This part of the development process relies heavily on the full development team for cost opportunity identification as well as implementation.

Design Rules Approach. Another method of minimizing design costs is to develop a series of design rules regarding very specific manufacturing processes. These rules essentially define the process capability and what needs to be reviewed if the process capability is violated. These rules have been written by many companies and can be purchased from independent sources. These rules define the requirements to achieve an effective process. Chip manufacturers have used design rules to insure that chips can be produced within the cost targets identified. If the process capabilities are followed, chip yields typically are very good.

Rules in the mechanical area have been published for years. For example, this includes types of bends which can be made given the equipment available, or how close to the edge of a bend a hole can be placed.

A third level of rules applies to the design process itself. One of the these rules was shown in Chapter 4, Figure 23, and is: Design to minimize parts and process steps.

Many of the generic rules appear to be common sense, but are not followed by designers, since during the process other problems draw their attention. In addition these are not rules in the true sense of the word, since exceptions can be found. Guidelines may be a better description. Here's a list of some of the more typical DFM rules or guidelines:

> Design to minimize parts/combine parts
> Design to minimize process steps
> Use gravity--design for downward assembly
> Use common parts wherever possible
> Eliminate product features not valued by the customer
> Eliminate fasteners/use self-fastening techniques
> Design out reorientations at assembly

Design out product adjustments
Design parts with easily identifiable features
Use self locating features like lead-ins
Standardize on fasteners
Design for modular assembly
Use geometric tolerancing
Minimize areas of tight tolerance/keep within low cost process capability
Size part designs within standard stock sizes, sheet sizes--optimize number per sheet

Design For Assembly. A common and well-liked method of DFM is to review the assembly and trace back the process from and assembly viewpoint. This method focuses the designer on minimizing the fastening steps, and on combining parts, thus minimizing the overall product costs. Design for assembly begins with the establishment of assembly and sub-assembly trees. By asking a series of questions, one can determine if parts are candidates for being combined. These questions are:

1) Is this part required to move relative to the other piece parts in the assembly?
2) Must it be made from a different material?
3) Must it be isolated either electrically or thermally?
4) Does it need to be disassembled when serviced?

If the answer to all of these question is no then the part should be considered for elimination or for combining with other parts. In a full DFA analysis every part will be tested for combining with others and identified as elimination candidates. The design can be scored as the number of parts eliminated divided by the potential number of parts to eliminate.

DFA is often referred to as the method of fasteners, where the method of fastening is examined in detail, with the idea being to simplify fasteners to the largest extent possible. DFA also gets referred to as the elimination of screws method.

Design For Test. This concept focuses the designer's efforts on the test operations with the goal of making test simple and easy to perform in a volume situation. Everything from TV's, computers and cars have self-testing processes designed into them. One of the most common methods is Built-In Self Test (BIST). This method requires the designing into the product the ability to carry out an explicit test on itself and report the errors if they occur. For this to be effective specific test hooks need to be placed into the hardware so that failures can be isolated correctly. The idea is to minimize the test time while maximizing the coverage. Each test should be constructed to identify problems down to the field replaceable unit. Each field replaceable unit can then be tested further after it's removed.

Test Options Influence DFT

The test engineer has several choices when it comes to the development of a test strategy. Each one results in different Design for Test requirements. Some of the options include:

Built-In Self Test. As previously discussed using this method of testing a diagnostic program is developed which is placed permanently into the product. Upon command the built-in self test program is activated, and tests the unit. It typically exercises the unit to identify problems. Built-in self tests are somewhat limited in their test technique. The test programs often run only high level functioning of the unit. One advantage of built-in self test is that it can be run every time the unit is powered on.

Boundary Scan Technology. This test method is built into the hardware design. Typically, this test method allows the

Developing Cost Sensitive Products

loading of a specified signal and the reading of signal responses at the device boundary. These outputs are compared to expected results for error detection. This testing is more accurate in pinpointing faults and is often used with Application Specific Integrated Circuits (ASICs) and Very Large-Scale Integration (VLSI) devices.

In-Circuit Testing. In-circuit testers provide an automated method to validate proper assembly and some component testing. These external testers can be programmed to test specific nodes within the board and check limited functionality. By changing the program, specific problems can be caught and screened from downstream manufacturing steps. Since this tester uses pin connections pushed against the printed wiring board surface, and since it needs to isolate specific circuits to be able to analyze and define faults, it requires specific considerations during the design process.

Shorts And Opens. This testing is often times part of the incircuit testing and is specifically oriented toward finding the process faults typically generated by automated and manual soldering techniques. Shorts and opens testers check for shorts among circuits and to the power planes, and unexpected opens. It tests through connection to the standard interface such as an edge connector.

Producibility

Producibility is usually understood as a check to insure a design can be manufactured, i.e., the company is able to produce it. However, in many companies, producibility has emerged into a much broader understanding and encompasses many more up front activities in the design process, including cost and manufacturing input, cost assessments, assembly time analysis, machine setup and time checks, cost reduction suggestions, vendor producibility assessments, mold analysis, raw stock and potential scrap analysis, new fixture minimization, optimization of flat stock usage to maximize output (i.e. fitting the maximum number of pieces per sheet of stock material), and much more. As a result in many companies the manufacturing function responsible for producibility has emerged as a primary interface for the designers. Producibility is a great way for designers to understand the capability of the current manufacturing processes, their limitations, and the design suggestions that drive product costs up.

Many producibility issues can be formulated into rules or guidelines that can be approved by a broader team. These rules can then be taught to designers so that they avoid unwanted conditions in future designs. One example of this is in electronic design where machine insertion space requirements can be defined so that board layout allows for the proper distances. In this case software exists which has incorporated these rules into automated checking. Another example is the listing of fixtures so that current fixtures can be utilized by designers where practical in order to avoid new fixturing costs and to minimize setup charges.

In today's world significant software products have emerged to help producibility and manufacturing engineers convert from raw stock to the product being designed. These tools allow trial and error electronically before machine time is used and fixtures or molds are produced, etc. These tools help the producibility engineer understand the cost implications in terms of time, setup, and scrap. Tools also exist for automated and manual assembly.

New Technology

This is an attractive and exciting area for designers. That is, to use the latest technology to achieve the cost and performance targets required by the product. However, this is a high area of caution since experienced designers will tell you that oftentimes with new technology only one of the three majors can be achieved. That is, either performance, cost , or quality can be achieved but not all three. The result is significant development delays, additional development costs, and possibly a product that won't work. Thus, any new technology areas need to be approached with reservations and the risks clearly understood.

Cost Pitfalls

Throughout the design process there are many cost pitfalls. Those are areas which lead to excessive cost being added to a product as it's developed. They are usually difficult for the team to avoid or eliminate entirely. These include:

a) Gross changes by senior management to the design late in the design process. This normally is very disruptive to the team process and hard to avoid if the senior team has not be involved early in the process.

b) The constant improvement in features and changing of the design. This has been called the rising cost of design enhancements. It is usually brought about by the lack of a clear schedule and a lack of a clear understanding of the customer needs. Once a concept is selected, new features become the domain of the next product.

c) Designing only part of the product and using other parts from previous designs. This normally leads to special problems at the interface of the parts which are avoided when an entire product is designed. Depending on the complexity of the interface, this can be a minor or a major cost.

d) Late changes in the product to account for new options, new market information, or failures during testing.

Ten Key Tips For Minimizing Costs. The following guidelines are ones that have proven useful over the years for minimizing costs. Even though they appear obvious, for many they are difficult to follow. Here's the list:

1) Minimize the number of parts
2) Minimize the number of process steps
3) Use common parts within the product and among products
4) Simplify both assembly and serviceability
   (Fasteners, assembly and disassembly times)
5) Maximize the designing out of non-value added operations such as in rework and inspection
   (Failure rate at test, repair times, extra handling, unreliable parts)
6) Design to leverage low cost overhead opportunities
   (Simplified configurations, preferred vendors, minimum number of vendors, etc.)
7) Design to minimize life cycle test costs
   (eliminate test redundancy)
8) Minimize tight tolerances
9) Integrate team concepts of design into the organizations
   (leverage full organization knowledge on new products)
10) Bring vendors into the development process as early as possible.

Summary

This chapter addressed the basics on developing cost sensitive products. Two areas are separated: the area of part selection and the area of part design. The techniques used for cost reduction are different for both. With regards to selections the following were covered: minimizing the features selected,

Developing Cost Sensitive Products

minimizing the cost of options, the use of volume components, use of industry standards and preferred vendors.

With regards to the design, a method was given that identifies a cost tool for use in cost trade-offs by designers. This is part of the Cost Driven Design approach. The Design for Manufacturing concepts were introduced with techniques for design analysis, process analysis, design for assembly, design rules, and technology analysis. The method of handling concept reviews was identified, and a listing of DFM rules was provided. Last, a few cost pitfalls were identified, and ten key tips for minimizing costs were given.

Chapter 8
Quality Focused Design

Product Quality

Quality in product design is what separates the good companies from the also-rans. Quality in product design is also what separates the good designers from the average engineer. Many an engineering team has introduced products that were technology leaders but did not have the quality level to sustain their initial momentum. Products that do not have an acceptable level of quality usually fail as a commercial product. Car companies have known this since losing a large part of their market to Japanese made cars. A quality product reasonably priced is what customers want.

What does this mean to designers? What should they consider when discussing the quality of their product and the quality of their design. This is a comprehensive topic, but simply put:

    1) Customers want their functional expectations met.
    2) Customers want the product to be free of real or perceived defects in workmanship or design.

Quality Focused Design 213

3) Customers want their products to last for very long periods.

All three are competitive differentiators for a company. Let's address each one.

Customers Want Their Functional Expectations Met

Earlier in this book we talked about several methods of trying to understand the customer inputs and relating it to the design process. These methodologies included QFD, Voice of the Customer, market research, focus groups, customer interviews, and telemarketing concepts. These methods help to collect and sort through the information and convert it into useful feature descriptions that can be used by design. These descriptions and the translation of them into product features need to be retested with the customers as the design is developed. This gives a higher degree of confidence that the final product will meet the customers expectations.

In addition to identifying the key features, it is also important to identify the customers perception of the quality of the product. Product features that look like they are easily broken, easily mishandled, or have areas of questionable quality will be less preferred in the long run. Design quality information will come out as models and prototypes are used in the product evaluation by customers. A good technique for evaluating the customer's perception of the quality of the product is to ask their judgment on certain features. These features can be rated as:

    Inadequate
    Adequate
    Well suited for the task
    Exceptional

CONSUMER REPORTS is excellent at rating the customer appeal of certain products. They rate products in 5 categories. These categories vary depending on the features being measured. In general they include:

> Much worse than average
> Slightly worse than average
> Average
> Slightly better than average
> Much better than average

CONSUMER REPORTS uses an expectation rating as a way of quantifying subjective measures but only where technical measurements don't exist. For example, in comparing an automobile the factual data would indicate miles per gallon, turning radius, engine size, transmission ratio, weight, head room, and luggage capacity. However, when trying to determine the customer preference and feature importance the above classifications would give a clear distinction to the designer on customer quality preferences. For instance, if the customer thought the car handling was "adequate" or "exceptional," or if the design of the dashboard provided a "much better than average" difference, the designer would get a better understanding of the customer's preferences. Much can be learned by trying to assess the customers expectations and the eventual implications on market share.

Customers Want Their Product To Be Free Of Real Or Perceived Defects In Workmanship Or Design

This is an obvious statement but very difficult to understand at the time of the design. Many historical studies have shown that low process quality leads to a higher incident of customer problems. That is, the more defects produced the more defects get through the inspection processes at the end of the line. The best way to control these defects is to develop designs that need little if any control, and that are not prone to error at assembly or fabrication. Characteristics of these designs are: simple to assemble, few parts, assembly fits are obvious and quick to do.

Testing of the product is also related to quality. The more rapidly and easily a comprehensive test can be performed the quicker a problem can be spotted and corrected. Thus, products with tailored test diagnostics can be evaluated quicker than those with multiple testing operations.

Recently (in the last 10 years) more formal methods to influence the design characteristic and relate them to downstream quality in manufacturing have developed. These include Taguchi's robust design, design of experiments, six sigma (which matches designed tolerance to process tolerance to minimize the potential of process variation causing defects). In addition, there are many DFM techniques that lead to fewer defects in production. These include the use of rules and methods. For instance, as parts are minimized the opportunity for error in the process is also minimized. Recent DFM related studies have shown that for assembly processes the amount of assembly time is inversely proportionate to the quality level of the part output. That is, low assembly time correlates with high quality output.

New Technology's Impact On Quality. This is a critical area that is often associated with high defect levels in manufacturing. These problems are associated with the immaturity of the process both internally and at the vendor. Quality problems add expense in the manufacturing process. This means more inspections, more testing, more repairs, and more time in the process. As a result more defects make it out the door and into the hands of the customer. New technology usage is a key area the development team must address during the design phase to control the quality risk to the product.

Cause And Effect Diagrams. One method of getting a better understanding of the effect of key design features on the manufacturing process is to develop a cause and effect diagram. This is a graphic method that is useful for illustrating the multiple causes of an outcome such as a defect. This method highlights not only the design characteristics that need to be controlled but also the process characteristics. Thus, the entire engineering community including design engineers, manufacturing engineers, and quality engineers can work together to control these potential defect contributors. A cause and effect diagram is shown in Figure 38.

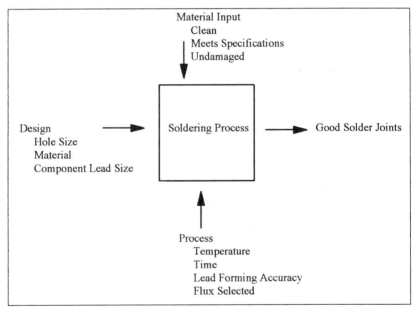

Figure 38 Cause and Effect Diagram. *These diagrams help in understanding the controls that need to be in place to achieve a good product. As can be seen the design choices have a major impact on these controls.*

Quality Focused Design

Opportunity For Defects. One method of understanding the chance for an error is the idea of opportunity for error. All the possible opportunity for errors are calculated and the designs with the lease opportunities are rated higher. This approach is good for simple designs or for problem parts, but as an overall method it can be misleading and too time consuming. The opportunity for defect calculation is different for assembly and fabrication. For assembly it can be calculated as follows. Each part has the potential opportunity to be picked up incorrectly (the wrong part), placed in the wrong orientation or position, attached incorrectly, and moved incorrectly to the next station. Each unique activity of part movement is counted as an opportunity for defects. Furthermore the previous process steps have opportunity for defects. For plastic parts the opportunity is equal to the number of mold sections which must be placed in the press, plus the number of materials, plus the number of process steps. Thus, for a snap fit plastic cover the opportunity for defects equals the number of opportunities at assembly and the number of opportunities at the fabrication stage.

Defect Analysis Approach. A very simple but direct approach to understanding defects is a review of the current defects being seen within manufacturing operations. Many times the defects that cannot be eliminated by process control appear as the top defects in manufacturing. Even defects like assembly errors can be analyzed and design changes put in place to correct these problems.

Customers Want Their Products To Last For Very Long Periods

This is well understood by manufacturers, and is a competitive advantage when the manufacturers are willing to provide a multiple year warranty for their products. Extended service contracts are also popular and provide advantages for many products. The customers want long periods for product life with only infrequent maintenance service. This period needs to be competitive within the industry. Fortunately, reliability on many functional parts can be calculated or estimated, and a combined reliability figure can be developed for products. Thus, the life estimate of a product can be defined and used as a guide in selecting components and in sub-assemblies. Many designers develop reliability budgets for each sub-system so that the summation of the parts will give the desired results.

Reliability is usually measured for complex designs. Simple designs generally don't require the same level of reliability management by designers. Reliability is usually stated in terms of Meant Time Between Failures (MTBF). This method identifies how long a product should last between expected failures.

Reliability Calculation. Statistical mathematical models are used for reliability calculations. Reliability is calculated as the probability of success. In other words, reliability is the probability that a product or system will perform as specified for a specific amount of time without failure. Reliability in complex assemblies is often stated using the exponential distribution. This can be given as

$$R(t) = e^{-t/u} \quad t \geq 0$$

R = Reliability stated as probability of successful operation for the time specified
t = Time the system is expected to operate without failure
u = MTBF

The calculation for the probability of the success of the system is the multiple of all the sub-system probabilities

Quality Focused Design

(assuming all the sub-systems need to operate simultaneously):

$$R_{System} = R_{Sub\text{-}system\ 1} \times R_{Sub\text{-}system2} \times ... \times R_{Sub\text{-}system\ N}$$

Thus, for a system to last for any reasonable length of time, each sub-system must have a reasonably high reliability. For example, electronic assemblies often follow this model since their components must operate simultaneously in order to produce the desired result. Take the example below of a simple electronic design which must operate for 1000 hours and is comprised of 8 parts where 6 of the 8 parts have a reliability of .98, sub-system 1 has a reliability of .97 and sub-system 2 has a reliability of .965. Then,

$$R_{System} = .97_{Sub\text{-}system\ 1} \times .965_{Sub\text{-}system2} \times .98^N{}_{Sub\text{-}system\ N}$$

where N = 6

Thus, $R_{System} = .829$

Faced with this low reliability problem, engineers can improve reliability by improving the reliability of each component. For example, if the objective is 94% reliability for the system, then each component must be 99% reliable. Additionally, one can lower the required operating time or build in redundancy to improve reliability. Note: Redundancy and parallel systems follow a different reliability mathematical model.

Thus, from the outset of design, the engineers should understand the reliability requirements. Using the mathematical model a reliability budget can be developed for each part. Reliability by itself is a complex topic, the summations of terms, the methods of calculation, and the probability distribution used vary depending on the product. The actual construction of a reliability model, the data collection and validation, and testing to support the data are very important steps in understanding the actual reliability numbers and in understanding the validity of these numbers.

Once reliability is understood and problem areas identified either a Fault Tree or Failure Mode and Effect Analysis can be used to determine the areas to target design changes. A fault tree traces symptoms back to the root cause. Failure Mode and Effect Analysis defines all possible failures and the effect on the system. Thus, it helps to focus in on catastrophic failures. The likelihood of failure must also be applied to both methods to understand the areas of concentration for design teams.

Regulatory Requirements

Regulatory requirements come in many forms. Governing bodies place regulations on products, industry groups develop standards and many corporations have their own standards. Oftentimes industrial and retail customers will specify that certain industry standards or regulations be achieved when developing products. Companies like GM, G E, and IBM all have their own standards that they impose and use to regulate incoming products.

Underwriters Laboratory (UL) is a widely recognized standards organization for safety. These standards are written for specific products depending on use, and focus on the safety of the user and repair people. Standards cover electrical and mechanical hazards, including electrical shock, dangerous access, flammability, and leakage current to name a few. A favorite UL test among engineers is the "finger test". It essentially says that a product cannot have openings large enough where a finger could accidentally be placed and be injured. Most products strive to meet these standards, and specific standards vary depending on product type. For instance:

UL 478 - Electronic Data Processing Equipment
UL 544 - Medical and Dental Equipment
UL 1244 - Measurement and Test Equipment
UL 1262 - Laboratory Equipment
UL 1469 - Telephone Equipment

Other organizations that are concerned about safety include IEC (International European Community), VDE (Germany), JIS (Japan), CSA (Canada).

Another type of standard is set by national and international organizations such as NEMA (National Electronic Manufacturers Association). This is a voluntary set of standards that equipment manufacturers have established to define the quality level of certain products. For example, NEMA standards have ratings depending on environmental requirements. One standard defines the requirement for an electronic enclosure that must operate in industrial environments where water may drip on the equipment. Another standard defines levels of acceptance for equipment that must operate stand-alone, outside in frigid conditions, sandstorms, desert areas, etc.

Government regulatory agencies have their own standards such as the Food and Drug Administration which regulates both products associated with human consumption and equipment that comes in contact with humans. The Federal Communication Commission regulates electromagnetic interference emitted from electronic equipment.

There is an entire host of regulatory agencies throughout the world. It is a well recognized way to insure product safety and quality in design. Because of the heavy regulations, it is very important for the development team to obtain the agency information since they affect many product design decisions. The result of meeting the standards and in doing the appropriate testing to validate that the new product meets the standard usually is a regulatory label that can be placed upon the product. Many purchasing departments and consumers look for these labels and their buying decisions are affected by this fact. These labels may not sell your product, but without them it will eliminate a major segment of any market.

Summary

This chapter discussed ways in which to understand quality at the early stages of design. It defined methodologies which help the designers understand if they are in fact meeting the expectations of their customers. Two methods of measuring the customer satisfaction level were identified so that designers can use these methods as a qualitative guide for product feature comparisons. In addition, this chapter discussed how to improve process quality at the time of design. It defined several design characteristics which improve process quality. The method of cause and effect diagrams was demonstrated, and the calculation of opportunity for defects defined.

Reliability of the product was introduced. An industry example showing the calculation of reliability was provided with formulas. Reliability as a competitive differentiator is an important one and the methodologies for understanding the sources of faults were identified.

Lastly, regulatory agency information was introduced and some descriptions regarding the specific standards needed for compliance were given.

Chapter 9
Development Time Management

Time To Market Approaches

The main differentiator of good design departments in the next decade will be their ability to develop competitive products extremely fast. Time will be the primary mode of competitiveness. Products that took 18 months to develop will need to be developed in 9 months, and those developed in 9 months will need to be developed in 3 months. Several leading professors argue that a 10x performance is needed in time reduction for companies to stay competitive in the global economy of the 21st century. The effect of time compression in the development cycle on profit can be seen in Figure 39.

This graph shows a typical product cycle. When spending occurs to develop the product and to start up manufacturing the cash flow is negative. Cash flow doesn't start to become positive until units are actually paid for by the customer. At this point the expenses of development should be dropping rapidly and the remaining expenses are those associated with the manufacture and minor maintenance of the product itself. Product income increases until the sales level out and the product reaches a steady volume.

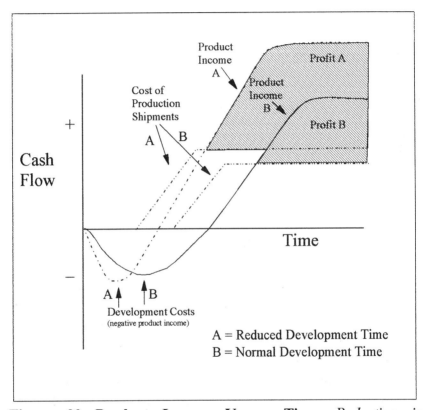

Figure 39 Product Income Versus Time. *Reduction in development time improves the profitability of the product. First the development costs are less. Second, since it is earlier to market there is less competition with the same features and the sales volume is greater. Third, it ships for a longer period of time, and often can sell for a higher price.* (Courtesy Prof. Dan Shunk Ph.D. Arizona State University)

Development Time Management

Two product cycles are graphed: one where the development cycle takes the normal amount of time (line B), and one where the development cycle is dramatically reduced by approximately 40% (line A). In a growing market the product developed on line A will grab a significant share of the new market given it has features that are not readily available elsewhere. In addition, its development costs are less and its manufacturing life is extended by several months. The net effect is significantly more profit.

Now if this same product is delayed in its shipping by 6 months due to development delays, not only are the development costs more, but the customers are not as attracted to it since other competitive solutions have started to emerge in the marketplace. The sales volume is lower than if it had been completed on time. Several market research studies have proven this product development phenomena.

Because of this, time is a very critical part of the design team life. There are many tools and methods used to deal with time when managing a small or large team, and with multiple dependencies on others not associated with the team. These techniques are discussed in this chapter.

Techniques For Project Time Management

PERT Charting. For a new team a good way to understand which activities affect one another is a concept called PERT (Project Estimating and Resource Tracking) Charting. This method is simply the defining and connecting of key tasks relating to development. The most common method to do this is to place key activities in circles (referred to as bubbles) and connect them to the next task with lines. Each bubble has a title, an owner, and a duration time and expected completion date. On a separate sheet of paper each bubble has a criteria for completion called the doneness criteria. This criteria is developed by the owner of the bubble and owners of dependency bubbles downstream and includes a list of needs by the other team members (see Figure 40).

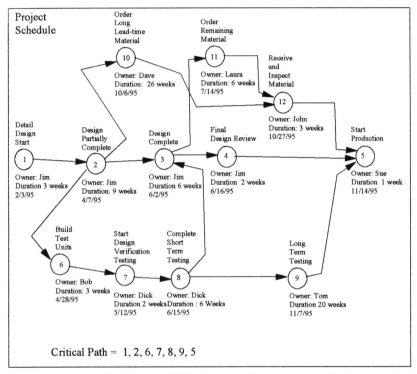

Figure 40 PERT Chart. *This is a well proven technique to help team members understand their part in the larger picture. It allows the correct scheduling of activities, and helps team members to understand which tasks impact the first product shipment date.*

When the PERT chart is developed each team member commits to developing his piece of the project within a certain time frame, given that the bubbles before them supply the needed information or activities as defined in the doneness criteria. Thus, the project leader can add up the time of all the tasks and understand how long the project will take, and identify the longest path of connecting bubbles. This is called the critical path. The project manager's main focus is then directed towards those items that affect the critical path, and in so doing keeps the project on schedule.

As changes occur during the program, dates can be updated and managed. Where time needs to be compressed, the project manager can ask for help from those outside the team and focus them on the time reduction of specific tasks. This may mean the adding of more people or simply the reduction of conservative estimates often referred to as "fudge factors". In Figure 40 if the project manager seeks an early decision on the test data, the project can be reduced by a few weeks.

PERT charting tends to be most effective on larger projects with multiple dependencies. PERT charting allows the project manager to raise awareness around schedule impacts in a constructive manner, and just as important de-emphasize those items that do not need to be rushed through. It is also effective when major changes to the development process are taking place, for instance the addition of new electronic tools, or the change of acceptance testing, or simply the addition of a QFD process.

PERT charts require significant updating to remain an effective tool since many changes occur after the start of the project. In order to do this effectively the person doing the PERT chart needs an automated tool to help adjust schedules, modify timing, etc. PERT charts can also be used to understand when certain development expenses are likely to take place and can be the basis of financial forecasts.

Milestones. Another good method of understanding time is a concept called milestone charting. The idea here is to map out the duration of specific tasks against the calendar. The start of the task is signified by a circle and the end of the task by a square. These circles and squares are referred to as the milestones of the project. Each milestone has an understood completion level. For example, a completion level called "design complete" means design drawings are complete and ready for review as agreed to by all team members. A milestone chart is a much simpler way to manage a project, and is very good for smaller programs or as summary schedules for major programs. If dependencies are already understood by the team and the project leader, then project tracking of tasks and budgets can easily be done using milestones (see Figure 41).

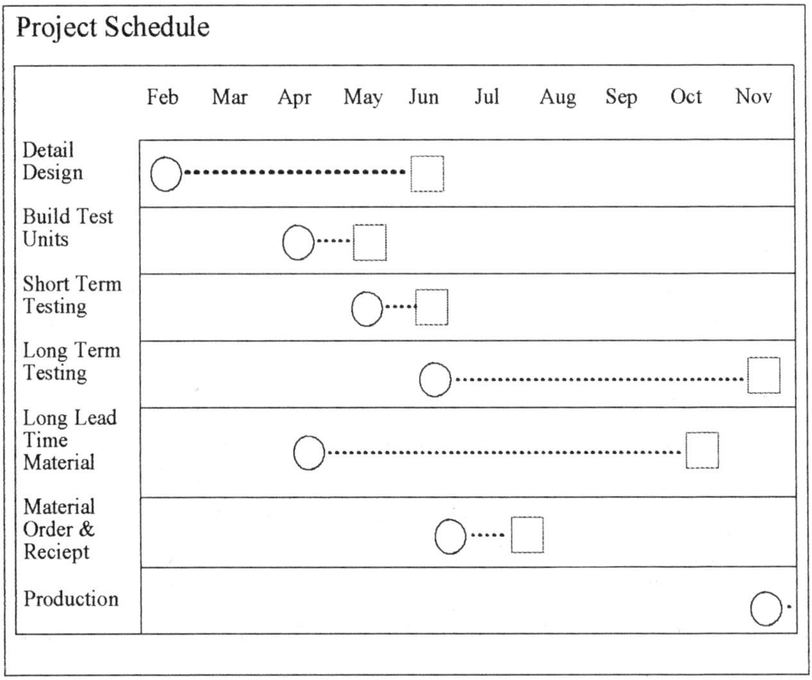

Figure 41 Milestone Chart. *Milestone charts quickly identify which tasks will consume the most time from the schedule. The items that have the longest lead times can be worked. The square is filled in when the task is complete.*

Development Time Management

Milestone charts also should be kept electronically, and each team member should have his own milestone charts that roll up into an overall team schedule. This schedule is easier to understand and can be provided to other project participants who are not on the direct team.

Schedules And Team Meetings

Monitoring schedules and team meetings is very important to keep the program on track. Each team meeting should include some of the current milestones with team members reporting their progress at the team meetings. Over the series of team meetings all the milestones not just the critical path items should be included in the discussions. The team meetings provide a forum for the team members to discuss the issues they are facing and elevate them to the team leader for help.

As discussed in Chapter 3, team meetings occur on a frequent basis, such as weekly or biweekly depending on the team's ability to get tasks completed. Each meeting is followed up with minutes and written action items for the team members and the team leader to resolve. Monthly, the schedule tool should be updated and reissued to the full team. Managers should get an updated schedule summary and tasks completed during the period. In this way the managers can help to keep the project on track without formal interaction with the team.

Ad Hoc Team Members. Each team has ad hoc members. These might be sub-assembly developers like power supply engineers or fluid system developers, or suppliers. Their activities affect the teams ability to stay on schedule. Since they significantly impact the ability of the team to perform they should attend specific team meetings prior to their deliverable due dates to report progress. In this way the entire team will understand the issues the ad hoc members are facing and be able to work to facilitate the resolution of any problems.

Expectation Definition

One helpful technique for managing the time aspects of the team meeting, and the project schedule is to define, at the earliest meetings the expectations of the project manager around communication. This is communication used at team meetings and between team meetings. For example, tasks should always be reported as status with regards to the milestone or PERT chart. For example, "we are 50% complete in our drawing development - 8 drawings out of 16 are done indicating we are on target to achieve the milestone date". The communication should not be "we are working on our drawings still and are on target to achieve the milestone date". More information must be passed than a simple reaffirmation of the committed date.

In order to simplify the discussion and improve the rapidness of the development cycle, problem reporting at the meeting should have a specific format. This format includes the following three items:

1) The individual has or does not have the information needed.
2) The individual or function is acting per the plan.
3) The individual or function is on or off schedule.

For instance, a problem might be reported as "I have all the information except two specifications. These were due to me last week. We are proceeding ahead but our due date is in jeopardy by an equal amount of time. In addition, one engineer was lost to another program and it could affect our ability to deliver on time. I'll give you a more accurate schedule impact next week".

Another example of a problem might be, "we had three failures in the test chamber last night. We are uncertain as to the cause. I need help getting the services of the thermal expert to help us characterize the probable causes. We have some time contingency in our plan for these delays but need to act quickly to resolve them".

Each communication presented the project manager with the three items need. In each case the project manager understood the type of problem, and the significance to the project schedule, and if the team member needed help to resolve the issue.

Keys For Very Rapid Development

The are some simple steps that can be taken to keep programs on track and occurring at a very rapid pace. They represent a 4 step process that if followed will help to keep the main tasks to the minimum time required. It is called DSCA (Define, Simplify, Control, Automate)

1) Define. Define everything and make sure its understood by the right people. Defining a problem is one of the quickest ways to bring about its resolution. Defining the schedule is a sure way to identify missing tasks, duplication of efforts, and processes that take far too long. For those lengthy processes define them in detail so that each step can be examined for possible elimination or replacement with a faster step. The simple act of defining helps to flush out the real issues and shorten the cycle of development.

2) Simplify. Simplify everything, including tasks, team meetings, process steps, designs, design effort, minutes, distribution lists, copies made, etc. Brainstorm during this process and look for creative ways to eliminate as much as possible. Can downstream testing be replaced by upstream simulation? Can prototype methods be used to make the initial production parts and thereby eliminate the vendor lead time from the critical path?

3) Control. Is each critical piece of the development effort under the control of someone on the team? Those items that are not under direct control are likely to be problematic. They will need to be worked by the project leader or delegated to a team member and monitored closely. If items are under no one's control, such as a software vendor updating his database product to work with your application program, you are probably in real schedule trouble.

4) Automate. Once the tasks have been defined, simplified, and under control, then automation is in order. For instance, CAD is automation, as is rapid prototyping. These items greatly simplify the development process. Information technologies should be used when they are understood and can contribute to the development process. Some of the key ones for development groups include client/server CAD, electronic mail, notes files, and other forms of electronic communication such as view and annotate tools for drawings. These tools help the team member to do his task effectively and efficiently.

Resolve Problems Quickly. Problems raised to the program manager must be resolved quickly in order to keep the team on track. Problems ignored, or swept under the table will develop an underlying feeling of stagnation and an inability to move the program forward. Problems resolved will facilitate the resolution of others and let the team members know they have help if the need arises.

Build Teamwork. Teamwork is the thing that separates new product development teams from simple collections of well intentioned people. Teamwork is the thing successful teams talk about the most. Steve Jobs, when he started Apple, emphasized the teamwork and team building he did at the start of the project. The teamwork concept reduced inhibitions and developed the trust needed for team members to share their ideas and contributions.

Break The Mold--Change. Look for areas to "break the mold". Find the quick ways of doing things--it may be unconventional, but find it. In a recent problem, one design team was quoted a three week production cycle for printed wiring boards. Instead, they went to a prototype shop and purchased the first three weeks' quantity deliverable in three days (at a slightly higher cost). Thus, they did not need to wait the lead time.

In another instance, one mechanical engineering group developed a method to send CAD files electronically to key suppliers and for the suppliers to be able to review and respond. Since each CAD file was transmitted an average of three times, and the changes affected other CAD files, the development cycle was reduced by 2 months. The previous method involved mailing files on 3 1/4 inch diskettes.

In another example, rapid prototyping was used at the beginning of the development process. It eliminated many of the problems that typically showed up late in the development cycle, thus, reducing the time of downstream processing. The list goes on and on, break your mold and innovate.

Innovate Off-Line First. Many an operational manager will want to see new products introduce new concepts. This often puts the new products at risk. Too many new implementations increases the probability of a schedule impact. Thus, as a rule, innovate off-line first before it is needed as part of the development cycle. Innovation is a key for successful products but it can bring unexpected problems, so prototype first before you depend upon it.

Concurrence Principle. As much as possible develop concurrently. Marketing and manufacturing should be developing activities in parallel. Their challenges are different than designs, but time consuming. Lead times for announcement activities can take several months, as can tool development and preparing for the process start-up. Thus, concurrency is an important issue. Factor it into the schedules, plans and meetings.

Critical Path. As much as possible let the most complex pieces be the critical path items and drive every other activity to never get on the critical path. The complex developments normally have senior management attention and have more weight behind them than the team can provide.

Timing In Project Reviews

In project reviews timing is always discussed. The project manager must clearly state those areas that are on schedule, behind schedule, or ahead of schedule. The remaining part of the review should be oriented towards the activities the team needs completed in order to avoid schedule delays. Oftentimes a recommendation to ship a limited configuration during the first two months or some other compromise position is needed to stay on track. Reviews are excellent for communication and receiving help, so direct the help needed by the team to solve the most serious issues.

Summary

In this chapter the effect of rapid time to market was discussed and charted. Two important techniques for time management were discussed in enough detail to be used. They were PERT charting, and milestone development. An appropriate methodology for monitoring schedules and handling schedule issues at team meetings was identified. The complexity problem of monitoring ad hoc team members was addressed. A methodology for communicating schedule impacts between team members was provided. Additionally, several key items needed to reduce development time were identified. One of these is a structured process called (DSCA) Define, Simplify, Control, and Automate. And lastly the method of handling the focus on timing in design reviews was briefly described.

Chapter 10
Fitting The Pieces Together

New processes, new product technologies, new software applications, new methodologies, new team structures, new measurements, and organizational change, represent significant innovation in the development process. Putting it all together and achieving success is difficult. Keeping it current and competitive is equally as difficult. Care must be taken to structure a process, develop responsibilities around that process, and develop the training and research to sustain the process in a constantly improving state. The previous chapters identified what needs to get done, why it needs to get done and how to do it. This chapter describes when to do various tasks and references previous chapters. It assumes that the steps mentioned as start-up steps in previous chapters have been completed (including budgeting, CEO support, functional supporting processes, etc.) and that the full organizational support structure is in place for implementation.

Fitting The Pieces Together 237

Steps to Success

In a normal operating CE environment the steps to a successful product can be outlined in a 4 step process. These steps were discussed in detail in chapter 3. It may be helpful to review them before following the process flow. These steps are:

Step 1) Starting the product
Step 2) Developing the detail
Step 3) Testing the product
Step 4) Launching the product

Starting The Product. This is the most significant step and includes establishing the team and the concept development. This step is shown in flow chart format in Figure 42 and Figure 43. In these figures the boxes in the middle show the steps being taken in the product development process. The boxes to the left identify the key tasks that need to be done. The boxes on the right identify the methodologies that occur and the tools used at these steps. As can be seen in Figure 42 the first steps in starting a product development activity are assigning the team, team investigation and training and the team's plan development. This is followed rapidly by concept development. These steps are shown in Figure 43 and include the development of the initial concept ideas, initial mock-ups, concept analysis, concept selection, and then review of that selection by senior management, team members, consultants and others and modification to the concept for concept finalization.

It should be noted that for specific products this generalized concept process may need to be tailored slightly to fit the uniqueness of specific products or business situations. The more problems that can be addressed and resolved at concept development the fewer the number of problems that will need to be addressed downstream. For instance, using Design for X (see Figure 43), where X is "regulatory requirements", new regulatory constraints can be placed at this level to avoid significant work in retrofitting the design with fixes during the detailed phase later in the process.

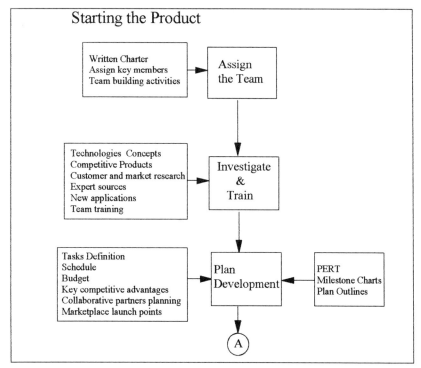

Figure 42 Product Kick-Off. *For a successful product kick-off the key team members are assigned, and the initial investigation and planning is accomplished. This flow assumes that the departmental environment exists to support the concurrent engineering concept and that the enabling information technology is in place to facilitate collaborative development. This includes any application programs needed for product development.*

Fitting The Pieces Together

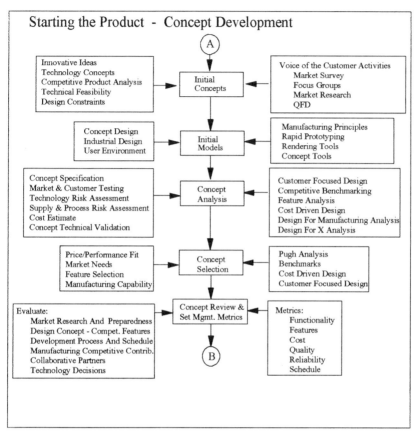

Figure 43 Product Concept Development. *In concept development several concepts and models are constructed and tested with customers. The final concept is either selected or is a combination of the best features of all the concepts developed. Before this step can be exited a thorough review of the work is needed by management to understand the market willingness to accept the product, the manufacturing capability to build the product, as well as the product cost, profitability considerations, scheduled implications, and competitive strengths of the product.*

Developing The Detail

Once the concept is chosen the detailed design work can begin. Almost simultaneously with the design development work the other work can be started (see Figure 44). The more collaborative the work effort between team members the better the design will be in addressing the issues.

As can be seen in Figure 44 the detailed design phase is very intense and requires excellence in communication. Many items need to be developed, analyzed, simulated, validated, checked for manufacturing effectiveness, etc. In Figure 44 the 5 boxes along the bottom, engineering analysis, simulations, parts library, regulatory analysis, and design for manufacturing are done concurrently as the design evolves. These analyses will take place whenever a sufficient part of the design is ready for analysis. The two boxes to the left, expert collaboration, and market planning will take place as required by the need. The boxes to the right, market and customer validation, metric validation, and rapid prototype validation take place when the design or part of the design is close to being completed, and are used to test the readiness of the design.

Automation of the communication, analysis, process etc. greatly speeds up this phase of the development cycle. Much must be considered during detailed development to have a good design. Additionally there are many trade-off decisions that need to get made, as well as conflicting pressures.

In addition to the design work, manufacturing is readying their processes, marketing is doing product testing with customers, and significant data, and preparatory work is taking place. At the end of this process is a formal design review by management. This design review covers 5 categories:

> Market research and preparedness
> Design development and project schedule

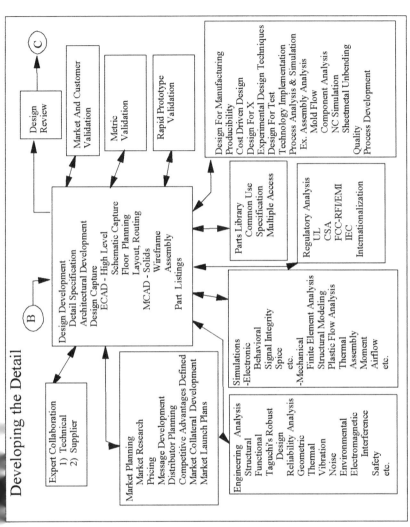

Figure 44 Detail Design Phase. *This figure shows the enormous complexity of the detail design phase of development.*

Manufacturing competitive contribution and readiness
Collaborative partners contribution and risk
Technology decisions, and implementation readiness

Further information on effective methods for design reviews was covered in Chapter 2 under "Product Development As A Structured Process".

Testing The Product. Once the detailed design phase is complete the product is ready for testing. This includes testing to insure the product operates as intended both in functionality and reliability. In addition this phase includes manufacturing readiness testing as demonstrated by pre-production builds, and marketing readiness through testing of the product with selected customers.

This phase starts with the development of a prototype or model which can be subjected to these tests. This phase of the development cycle is shown in Figure 45. The design is still fairly dynamic during this phase but at the detailed level. Exiting this phase the design should stabilize, with the intent being that all the bugs are out of the product. During this phase select customers get to test the product, provide their feedback, and hopefully are delighted with the design.

This phase ends when sufficient testing is completed to know the product functions as designed, and can handle the rigors of manufacturing, shipping and customer use. In addition, customers have seen the product, and usually they have had prototype product to test or validate and have provided their feedback. For marketplace success these customers should be anxiously awaiting the delivery of the first production quality product and delighted with the new features and functionality.

The final design review for this phase covers the following 5 activities:

Market test results, messages used
Design testing results and corrections

Fitting The Pieces Together 243

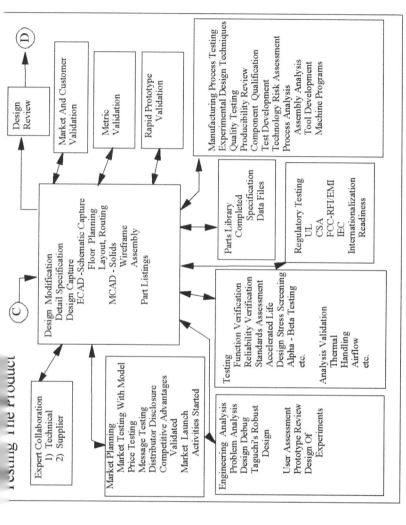

Figure 45 Testing The Product. *This figure shows the continued complexity of the development process as it moves throughout the design testing phase.*

Manufacturing process validation
Collaborative partners readiness
Technology preparedness and qualification

Launching The Product. After the testing has reached a level where little change is anticipated in the design, then formal launch activities can accelerate. The functional organizations become significantly more involved in preparation for full ownership and in placing the design into full operational mode. The development focus shifts to initializing business operations. This phase is typically very short for the new product team. It can be seen in Figure 46.

Much of the activity in this phase is education and finalization of the work done during the development process and documenting this work for future reference by other developers or those who will maintain the product during its life.

The final design review of the product is really the kick-off review for the full marketing, distribution and manufacturing functions. It covers the same categories as previous reviews, but here the work must be completed to consider it as closure for the design team. Marketing plans are reviewed in detail as are manufacturing readiness in operations, material availability, and quality metrics. Sales and manufacturing training are planned.

It should be noted that FRS (First Revenue Shipment) can occur before this phase is completed, and often occurs in the previous phase, when sufficient testing is completed to know the product will function correctly.

Some products warrant further work for cost reduction purposes after the product is launched. This is often caused by technology availability reasons and represents major change to a product when the new technology is factored into the product.

In addition, collaborative efforts with partners on product marketing and on sales and distribution are launched.

Fitting The Pieces Together 245

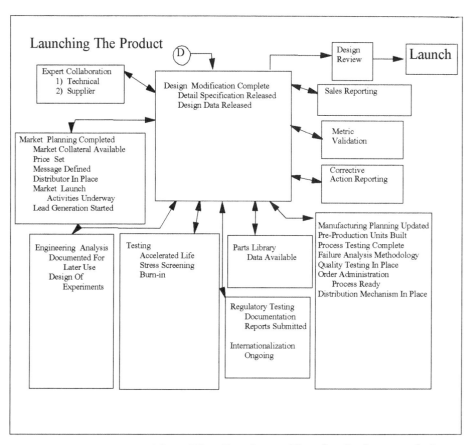

Figure 46 Launching The Product. *The final phases of the development process includes the hand-off to manufacturing and marketing operations. A substantial amount of education is needed for a successful hand-off, as well as communication literature. The number of people who become involved with the product grows enormously. During this phase design is supporting manufacturing and marketing efforts.*

Summary

In this chapter the various elements of the concurrent engineering process were tied together in an overall process flow. This built upon the descriptions of the process as defined in Chapter 3. The main elements of the process were connected and shown in a detail flow diagram. These elements are:

Step 1) Starting the product
Step 2) Developing the detail
Step 3) Testing the product
Step 4) Launching the product

Concurrent engineering occurs as simultaneous development and is not a sequential flow. This parallel development is shown in the figures in this chapter.

References

This is a listing of the sources and works consulted by the author or which influenced the author's thinking. This list is offered for those who wish to pursue the subject of concurrent engineering in greater detail.

Akao, Yoji, *Quality Function Deployment: Integrating Customer Requirements into Product Design*, Productivity Press, Cambridge, MA.
Barkan, Philip, Simultaneous Engineering, *Global Competition - Breaking the Barriers with Concurrent Engineering* (Conference Notes), October 24, 1990, Digital Equipment Corporation, Marlborough, MA.
Barkan, Philip, Some Impediments to Successful Implementation of Simultaneous Engineering. *Global Competition - Breaking the Barriers with Concurrent Engineering* (Conference Notes), October 25, 1990, Digital Equipment Corporation, Marlborough, MA.
Carter, Donald, and Barbara Baker, CE Concurrent Engineering: The Product Development Environment For the 1990's, Addison-Wesley Publishing Company, Reading, MA
Clausing, D.P. Improved Total Development Process - *Changing The 10 Cash Drains Into Cash Flow In Quality Function Deployment,* American Supplier Institute, Dearborn, MI, 1987.

Dehnad, Khosrow, *Quality Control, Robust Design, and the Taguchi Method*, Chapman and Hall Publishing, New York, New York, 1989.

Gillen, Dennis, and Elizabeth Fitzgerald, Expanding Knowledge And Converging Functions, *Concurrent Engineering*, May/June 1991, Auerbach Publishers, New York, pp. 20-28.

Henderson, Mark, and Dan Shunk, Small and Medium Enterprise Competitiveness in a Global Economy, *Circle of Excellence*, (Conference Notes), November 1993, Digital Equipment Corporation, Marlborough, MA.

Hill, Sidney, Modeling Techniques Spark Fierce Competition in CAD/CAM, *Manufacturing Systems*, September 1994, Chilton Publications, Carol Stream, IL.

Hinckley, Martin, A Global Conformance Quality Model, *Autofact'93 Abstracts* (*Conference* Notes), November 9, 1993, SME Press, Dearborn, MI.

Keller, Eric, Client-Server CAD, *Concurrent Engineering*, May/June 1991, Auerbach Publishers, New York, pp. 44-46.

Kempher, Lisa, Profile of Excellence, *Computer-Aided Engineering*, October 1993, Penton Publishing, Cleveland, OH.

Kurland, Raymond, Understanding Variable-Driven Modeling, *Computer-Aided Engineering*, January 1994, Penton Publishing, Cleveland, OH.

Neff, Robert, A QFD Snapshot. *Business Week Special Issue, The Quality Imperative*, October 25, 1991, McGraw-Hill, New York, pp. 22-23.

Parikh, Yogesh, Parameter Design For Product Optimization, *Excellence*, November 1986, No 166, Digital Press, Digital Equipment Corporation, Marlborough, MA.

Parikh, Yogesh, Engineering For Quality By Design - The Taguchi Approach, *Excellence*, September/October 1986,

References 249

Digital Press, Digital Equipment Corporation, Marlborough, MA.

Power, Christopher, Walecia Konrad, Alice Cuneo, and James Treece, Value Marketing, *Business Week*, November 11, 1991, McGraw-Hill, New York pp. 132-140.

Pugh, Stuart, *Total Design Integrated Methods for Successful Product Engineering*, Addison-Wesley Publishing Company, Reading, MA.

Pyzdek, Thomas, *What Every Engineer Should Know About Quality Control*, 1989, Marcel Dekker, New York.

Salomone, Thomas, Reducing Product Cost & Development Time *Circle of Excellence* (Conference Notes), November 1993, Digital Equipment Corporation, Marlborough, MA.

Schmidt, Vern, Rapid Technologies for Customer and Supply Chain, *Autofact '93 Abstracts* (*Conference* Notes), November 9, 1993, SME Press, Dearborn, MI.

Sprow, Eugene, Chrysler's Concurrent Engineering Challenge, *Manufacturing Engineering*, April 1992, SME, Dearborn, MI.

Taguchi, Genichi, & Phadke, Madhav, Quality Engineering Through Design Optimization, *Conference Record, GLOBECOM84 Meeting*, IEEE Communications Society, Atlanta, GA. pp. 1106-1113.

Welter, Therese, How To Build and Operate a Product Design Team, *Industry Week*, April 16, 1990 pp. 35-50.

Whiting, Rick, Managing Product Development From The Top, *Electronic Business*, June 17, 1991, The Cahners Publishing Co., Newton, MA. pp. 40-44

The Institute for Defense Analysis IDA Report R-338

Index

Activity based accounting, 130
Agendas, 89
Aggressive growth, 188
Akao, Yoji, 102, 108
Announcement activities, 78
Anticipated cost solution, 124, 125
Application programs, 147
Assembly analysis, 158
Associativity, 151
Athletic vs. industrial team, 64–66
Automated device selection, 161

Award-winning design, 9

B-2 Stealth bomber, 23
Baseline features, 144
Behavioral factors, 4
Behavioral schematic capture, 160
Benchmarking, 106, 109, 140
Best practices survey team, 23
Bill of materials, 196
Board layout, 162
Boeing 777, 24
Boundary scan technology, 207

Brainstorming, 89, 91
Branching methodology, 194
Bubble, 225, 226
Budgeted dollars, 200
Budgeting guidelines, 34
Built-in, self-test, 206

Cause and effect diagrams, 216
Changing market, 19
Client/server CAD, 10, 172–175, 232
Client/server groupware, 11
Clients, 172
Collaboration, 7, 55, 169, 173
  behavior, 8, 12, 28
  development, 14, 22, 238
  elements, 28
  environment for, 16, 88
  partners, 240
  within process, 46
  relationships, 25
  team, 70
Collocation, 91
Communication network, 10
Communication software, 147
Communication technology, 9
Comparison matrix, 119–120, 142
Competitive advantage, 10, 114, 116, 120, 124
Competitive analysis, 82, 83, 106, 140, 186
Competitive teardown, 76, 141

Competitor's pricing, 109
Computer aided design, 3, 147–149, 172
Computer aided engineering, 3, 23, 148, 172
Computer networking, 10
Concept
  datum, 133
  development, 37, 238
  reviews, 202, 239
Concurrence principle, 233
Concurrent engineering (CE), 2, 15, 23, 37
  concept, 7, 15, 176
  definitions, 22, 25
  developing the detail, 240–241
  difficulties, 144
  environment for, 9, 154, 237
  fundamentals, 7
  implementation, 33, 236
  launching the product, 244–245
  major initiative, 17
  mentor, 23, 80–81
  process, 16, 28, 29, 100, 236
  process flow chart, 12, 236–237
  starting the product, 237–239
  team, 169
  testing the product, 242–243
Constraints, 152

Index

Consultants, 8
Convergence of ideas, 12–13
Core team members, 96, 98
Corporate culture, 21
Cost accounting impact, 130
Cost budget, 126
Cost Driven Design, 102–103, 115, 197–199, 210
  competitive comparison, 119–120
  cost solutions testing, 120–121
  cost target analysis, 116, 117, 120, 123, 126, 142
  cost trade-off analysis, 200
  determination, 38
  example, 115, 117–119, 121–126
  flow chart analysis, 203
  modification analysis, 118, 124
  tracking matrix, 122
Cost effective features, 180
Costing tools, 126
Cost management, 197
Cost pitfalls, 209
Cost reviews, 202
Cost-sensitive products, 191
Cost solutions testing, 120–121
Cost target analysis, 116, 117, 120, 123, 126, 142
Cost trade-off analysis, 200
Critical features, 78
Critical path, 226–226, 234
Customer delighters, 14, 107, 192

Customer demands, 110–111, 113
Customer feedback, 8, 181
Customer Focused Design, 106, 143
Customer satisfaction, 19, 109, 114

Data libraries, 10
Database management, 164, 166
Database tools, 127
Datum, product, 132–133
Defect analysis approach, 217
Design
  analysis, 201
  cycle compression, 6
  enabling technology, 43, 169
  engineering's role, 75
  influence on costs, 36
  optimization tools, 157
  planning, 85
  reviews, 85, 241
  rules, 204
  stress analysis, 138
Design for Assembly, 126–127, 205
Design-build teams, 24
Design of experiments, 23, 139, 215
Design for Manufacturing, 4, 45, 102, 115, 200, 205, 215
  concept stage, 115
  detailed design stage, 126
  end of cycle, 128

[Design for Manufacturing]
  guidelines approach, 127
  rule of minimization,
    128–129
  rules approach, 127, 205
Design for Test, 126, 206
Design for X, 103, 131
Detailed schematic, 161
Determined cost, 38
Developing details, 45, 51,
  240–241
Development costs, 38
Development cycles, 19
Development time shift, 16
Development process and CE,
  39–41
Device libraries, 161
Distributed team members, 91
Distribution channel
  development, 14, 179
Distributors, 8
Doneness criteria, 225
Dynamic analysis, 148, 156

Early involvement, 70
Efficiency in design, 26
Electronic assembly CAD,
  157
Electronic communications, 3
Electronic design, 9
  tools, 160
Electronic mail, 3, 148, 163
Enabling technology, 17, 149,
  176
Engineering change orders,
  23, 164

Engineering department
  efficiency, 166
  functional support, 84
  services, 84, 96
Engineering team leader, 96
Expectation definition, 230
Experimental design techniques, 105, 136, 138
Explicit modeling, 152

Feature based modeling, 152,
  154
Feature libraries, 154
Feature selection, 192
Finite element analysis, 148,
  155
First revenue shipment, 244
Fixture database, 160
Flow chart analysis, 203
Focus groups, 181, 187
Forming teams, 62
Fudge factors, 227
Functional characteristics,
  108, 111, 114
Functional inputs, 44
Functional preparedness, 57
Fundamental change, 32

Geographical region, 183
Global communications, 165
Groupware tools, 163

Hole alignment, 150
Human factors, 9

Importance ratings, 111

In-circuit testing, 207
Incurred costs, 38
Industrial teams, 62, 69
Information architecture strategy, 168–172
Information library, 156
Information technology, 3, 7, 9, 29, 56, 96, 169
Institute for Defense Analysis, 22
Integrated product definition team, 23
Interface design, 195
Interfunctional commonality, 86–87
Islands of automation, 10

Knowledge of design, 38

Launching the product, 244–245
Lead engineer, 96
Life-cycle costs, 33, 38
Logic optimization, 161
Loss equation, 136–137

Machining simulation, 148
Make/buy decisions, 197
Management
 limitations, 93–94
 stoppages, 58
Manufacturing department
 competitive analysis, 76, 82
 engineering tools, 157
 functional support, 82–83
 guidelines, 126

[Manufacturing department]
 infrastructure, 82–83
 input, 2, 76
 process development, 14, 83, 239
 process files, 159
 process generation, 162
 team member, 78
Market, 179
 competitive assessment, 43, 186
 positioning, 81, 187, 188
 potential, 183, 185
 segmentation, 183
Marketing department
 functional support, 81
 input, 2
 manager, 96
 process development, 14
 role, 77
 tools, 162
Market share, 109, 114
Master file, 157
Matrix management, 67
McDonnell Douglas Aerospace, 23
Mechanical CAD (MCAD), 154
Mean time between failure, 131, 218
Mechanical designs, 9
Methodologies, 3, 29, 100, 101, 107, 144, 236
Milestones, 227–228
Minimizing costs, 210
Minutes, meeting, 89

Modification analysis matrix, 118, 124
Mold flow analysis, 126
Mold simulation, 148, 158

Nationwide network, 23
National Electronic Manufacturers Association (NEMA), 221
Networks, 23, 147, 167–168
New technology analysis, 208
Noise factors, 134
Noise variation, 135
Non-collaborative behavior, 8
Northrop Corporation, 23
Notes files, 163
Numerical control (NC) machines, 148, 158

Opportunity for defects, 217
Option cost analysis table, 196
Options selection, 193
Organization
  boundaries, 4
  segmentation, 4
  structures, 28
  support, 28
  trust, 6
Orthogonal array, 139
Over-the-wall syndrome, 4–5, 66–67
Overhead costs, 130

Parametric design, 104, 153
  modeling, 152–153
Parts library, 155

Pattern testing, 159
Performance benchmarking, 140
Performance ratings, 124
Personal computers (PCs), 172
PERT charting, 225–227
Phased development process, 49–50
Plastic analysis, 148, 158
Pricing, 188
Process
  analysis, 201
  parameters, 139
  simulation, 159
  technology, 7, 41, 119
Producibility, 5, 207–208
Product, 38
  cost, 20–21, 38, 191
  data management, 148, 164
  design review, 86
  functionality, 3
  goals, 49, 61
  income vs. time, 53, 55, 224
  launching, 46, 52, 244–245
  life cycle, 35
  profitability, 38, 55
  quality, 19, 212
  samples, 182
  support, 96
Profit, 20, 38, 53, 55, 224
Project
  management, 57
  reviews, 234
Pugh process, 59, 104, 132–133

Index

Quality
  loss, 134-137
  QC analysis, 160
Quality Focused Design, 212
Quality Functional Deployment, 4, 57, 101-102, 108, 112, 145, 185, 213
  example, 108-114

Rapid product development, 54, 231-232
Rapid prototyping, 10, 83, 106, 141, 143, 149, 159
References, 247
Regional market, 179, 184
Regulatory requirements, 220
Reliability, 105, 131, 218-222
Rending tool, 156
Research infrastructure, 83
Resellers, 8, 14, 76
Resource assessment, 59, 60
Return on development, 53, 224
Reverse engineering, 141
Robust design, 104, 134, 136
Rule of tens, 36

Segmented organization, 6
Senior manager role, 21-22, 72
Sequential design process, 2, 4, 16, 39, 73
Setting market requirements, 75-76
Sheet metal unbending, 158

SIC codes, 183-184
Signal integrity, 162, 201
Simulation, 9, 161
  of mold flow, 126
  tools, 79
Simultaneous development, 24
Six sigma, 215
Skill vs. role confusion, 92
Software strategy, 167
Solids modeling, 131, 148, 150, 151
Space Station Freedom, 23
Spending rate, 35
Spread sheet, 156
Standard features, 154
Starting the product, 44, 51, 237-239
Statistical Process Control (SPC), 136
Stereolithography, 141, 143
Strategic design plan, 44
Structural simulations, 148
Structured design process, 27, 49
  methodologies, 3, 14-15, 57
Subfunctional analysis, 194
Supplier, 10, 195
  alliances, 8, 79
  input, 70
Support team, 79-80
Surfacing, 150

Taguchi method, 135
  quality loss curve, 137

[Taguchi method]
  robust design, 104, 134, 136, 215
Targeted customer, 180, 183
Transaction Control Processing Internet Protocol (TCPIP), 167, 168
Team
  assigned, 73
  autonomy, 84
  CE, 68–71
  collaboration, 8, 68–69, 85
  difficulties, 88
  example of, 95–96
  functional approach, 68–69
  goals, 85
  leadership, 73
  limitations, 92, 94, 95
  management, 84
  matrix approach, 67, 69
  meetings, 89, 229
  membership, 59, 80, 229
  participation, 74
  plan, 74
  representative approach, 68
  teamwork, 8, 232
Technology curve, 2
Telemarketing, 181
Testing the product, 242–243
Thermal analysis, 157
3-D modeling, 3, 150–151

Three T's of CE, 60
Time management, 223, 225
Time to market, 2, 223
Tolerance
  analysis, 158
  design, 104
  modeling, 148
  tightening, 105
Tools, 147, 176
  limitations, 175
Total quality management, 23
Tracking matrix, 122–123
Trial number, 138
Tuning, 168

Umbrella strategy, 17
Underwriters laboratory, 220

Value added features, 144
Variable driven modeling, 151
Variational modeling, 152–153
Viper, 15, 188–189
Visualizations, 44, 156
Voice of the customer, 213

Wireframe, 150
Workflow, 148, 164–165, 174–175
Workstations, 172

About The Author

THOMAS A. SALOMONE is Manager of the Design Business Group at Digital Equipment Corporation, Maynard, Massachusetts. The author of several professional papers on concurrent engineering and new management practices, he is a member of the Society of Manufacturing Engineers and the American Society of Mechanical Engineers. Mr. Salomone, a consultant to many companies on the design and implementation of concurrent engineering, received the B.S. degree (1971) in physics from Fairfield University, Connecticut, and the M.S. degree (1976) in administration and management from Union College, Schenectady, New York.